Concrete Pressure Pipe

Acknowledgment

This completely rewritten version of AWWA Manual M9, *Concrete Pressure Pipe,* was prepared by the Standards Committee on Concrete Pressure Pipe. The membership of this committee at the time it approved the current manual was:

Walter K. Neubauer, *Chairman*
Conrad Hohener, Jr., *Vice-Chairman*
Joseph A. Willett, *Secretary*

Members

Cliff J. Arch	Gerald G. Emerson
Tom Aulita	John O. Grimsley
James H. Bailey	Lewis Keyser
Robert E. Bald	Enn Kiilaspea
Charles E. Beal	A.C. Michael
William E. Bradbury	Albert McDonald
Wayne Brunzell	Charles A. Parthum
Duane D. Buchholz	Matthew W. Piche
Stanley M. Dore	Alexander E. Scalzitti
Stephen E. Dore, Jr.	Robert A. Skinner
Thomas C. Earl	E.L. Wright

©Copyright 1979 by the American Water Works Association
Printed in US

CHAPTER 2

Description of Concrete Pressure Pipe

General

Several types of concrete pressure pipe are manufactured and used in the US and Canada. Because of their construction, some pipes are made for a specific type of service condition; other pipes are constructed in such a manner that they are suitable for a broader range of service conditions. The description of these pipes, in general terminology as in Table 2.1, is based on whether or not the pipe has a full-length steel cylinder and whether it is conventionally reinforced (that is, with reinforcing bar, wire, or smooth bar) or prestressed (that is, with high-strength wire wrapping). It should be noted that not all concrete pressure pipe manufacturers make all of these types of pipe.

Prestressed Concrete Cylinder Pipe (C301)

Prestressed concrete cylinder pipe has been manufactured in the US since 1942. It is the most widely used type of concrete pressure pipe. Some of the applications for this product are transmission mains, distribution feeder mains, water intake and discharge lines, pressure siphons, penstocks, industrial pressure lines (including power plant cooling-water lines), sewer force mains, gravity sewer lines, subaquaeous lines (into both fresh water and salt water), and spillway conduits.

Types of construction. Prestressed concrete cylinder pipe has two general types of construction: (1) pipe with a steel cylinder lined with a concrete core, and (2) pipe with a steel cylinder embedded in a concrete core (see Fig. 2.1). In either type of

Fabrication of a steel cylinder

TABLE 2.1
General Description of Concrete Pressure Pipe

Type of Pipe	AWWA Spec.	Steel Cyl.	Non-Cyl.	Mild Reinforcing Bar	High Strength Wire	Design Basis*
Reinforced concrete cylinder pipe	C300	X		X		Rigid
Prestressed concrete cyl. pipe	C301	X			X	Rigid
Reinforced concrete non-cyl. pipe	C302		X	X		Rigid
Pretensioned concrete cylinder pipe	C303	X		X		Semi-rigid
Prestressed non-cyl. pipe	None		X		X	Rigid

*The terms "rigid" and "semirigid" in this table and in the remainder of this manual are intended to differentiate between two design theories. Rigid pipe does not depend upon the passive resistance of the soil adjacent to the pipe for support of vertical loads; semirigid pipe does require this passive soil resistance. The terms "rigid" and "semirigid," as used here, should not be confused with the definitions stated by Marston in Iowa State Experiment Station Bulletin No. 96.

construction, manufacturing begins with a full-length welded steel cylinder. After joint rings are attached to each end, the pipe is hydrostatically tested to ensure water tightness. A concrete core with a minimum thickness of 1/16th times the pipe diameter is placed either by the centrifugal process (lined cylinder pipe) or by vertical casting (embedded cylinder pipe). After the core is cured by steam or water, the pipe is helically wrapped with high strength hard-drawn wire using a stress of 75 percent of the minimum specified tensile strength. The wrapping stress ranges between 150,000 and 220,000 psi. The wire spacing is accurately controlled to produce a predetermined residual compression in the concrete core. The core is covered with a cement slurry closely followed by a dense coating of mortar or concrete that is rich in cement content. Steam or water is used to cure the coating.

Size range. Lined cylinder pipe is available in diameters of 16-60 in. Embedded cylinder pipe is commonly available in diameters of 24-144 in. Pipe larger than 200 in. in diameter has been manufactured. Lengths are generally in the 16-24 ft range, although longer units can be furnished depending on the weight and dimensional limitations of shipping.

Design basis. Prestressed concrete cylinder pipe has been designed for operating pressures greater than 400 psi and earth covers in excess of 100 ft. The design of this pipe is covered by either of two design methods listed as Appendices A and B in AWWA C301. Both of these design methods are based on combined loading conditions, which is the most critical type of loading for rigid pipe, and include surge pressures and live loads. Generally, Appendix A is used by engineers and specifying agencies in the eastern half of the US and Canada. Appendix B is used by those in the western half of the US.

Joints. The standard joint for prestressed concrete cylinder pipe is sealed with a rubber gasket and steel joint rings, as shown in Fig. 2.1. The spigot ring is a rolled shape containing a rectangular recess for holding a continuous solid ring gasket of circular cross section. The gasket is compressed by the flat portion of the steel bell

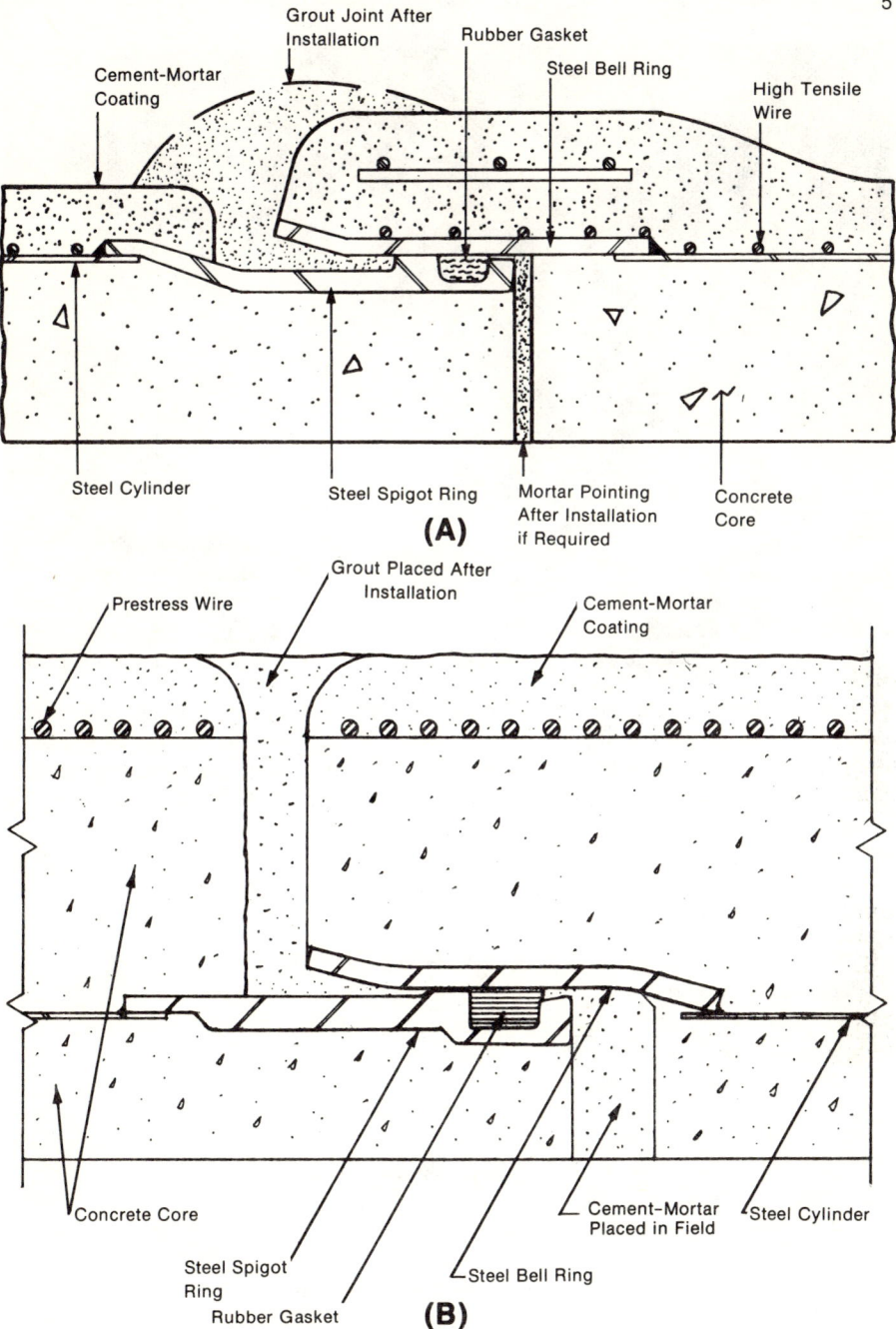

Fig. 2.1 Prestressed concrete cylinder pipe
(A) Typical lined cylinder pipe cross section.
(B) Typical embedded cylinder pipe cross section.
NOTE: External joint grout poured in diaper furnished by pipe manufacturer.

Joint ring being welded to steel cylinder, which is typical of C300, C301, and C303 pipe

ring when the spigot is pushed into the bell. Both joint rings are sized to close tolerances on a hydraulic press by expanding beyond the elastic limit. Core diameter and volume of the gasket are tightly controlled to ensure a reliable high pressure seal. After field assembly, the exterior joint recess is grouted to protect the steel joint rings. Other methods of protecting the joint rings are available. The internal joint may or may not be pointed with stiff mortar depending on the type of water the pipe will carry and the type of protective coating applied to the joint rings during manufacture.

Reinforced Concrete Cylinder Pipe (C300)

Prior to the introduction of prestressed concrete cylinder pipe (C301) in the early 1940's, most of the concrete pressure pipe in the US was reinforced concrete cylinder pipe. As the names would indicate, the construction of this pipe is similar to prestressed concrete cylinder pipe except that mild steel reinforcing is cast into the wall of the pipe instead of prestressing with high strength wire. New installations of reinforced concrete cylinder pipe have been declining over the years as prestressed concrete cylinder pipe has gained wide acceptance based on its excellent performance and lower cost.

Manufacture. Manufacture of reinforced concrete cylinder pipe begins with a hydrostatically tested steel cylinder and attached steel joint rings. The cylinder assembly and one or more reinforcing cages are positioned between inside and outside forms, and the concrete is placed by vertical casting. Steam or water is used to cure the concrete

Size range. Reinforced concrete cylinder pipe is manufactured in diameters of 24-144 in., with larger sizes limited only by the restrictions of transportation to the

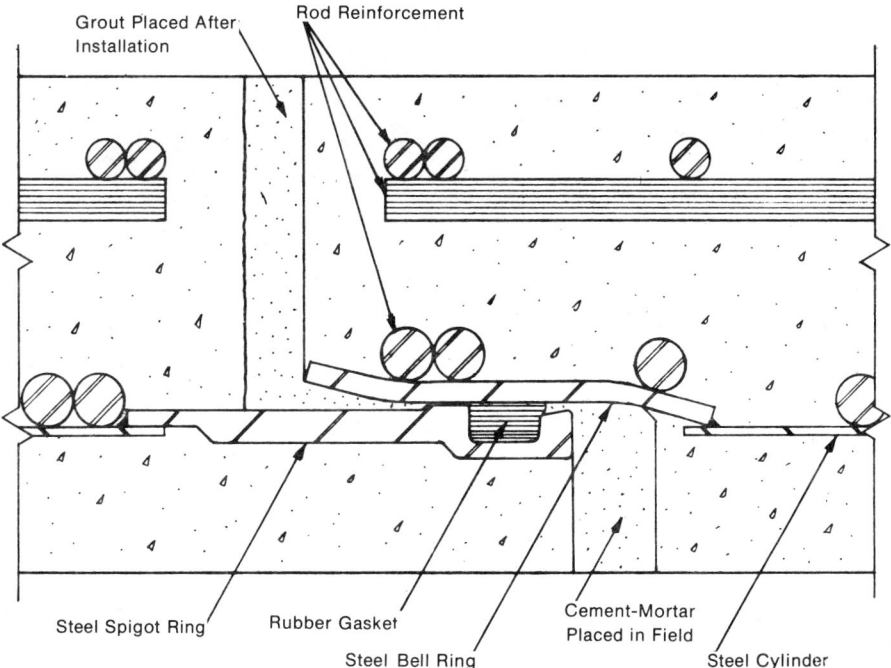

Fig. 2.2 Reinforced concrete cylinder pipe

job site. Standard lengths are in the 12-24 ft range. The applications for this type of pipe are the same as for prestressed cylinder pipe.

Design basis. Design of reinforced cylinder pipe is covered by a design appendix to the C300 specification, which includes external loads and internal pressures individually and in combinations. The reinforcing steel furnished in the cage(s) is no less than 40 percent of the total reinforcing steel in the pipe. Although the maximum loads and pressures for this type of pipe depend on the pipe diameter, wall thickness, and strength limitations of the concrete and steel, it is uncommon for this pipe to be used at pressures greater than 250 psi or in trench installations exceeding 20 ft of earth cover.

Joints. The standard joint for reinforced cylinder pipe, as shown in Fig. 2.2, is identical to the joint for prestressed cylinder pipe, consisting of steel spigot and bell rings and a rubber gasket. As with C301 pipe, the external joint recess is normally grouted and the internal joint space may or may not be pointed with mortar.

Reinforced Concrete Noncylinder Pipe (C302)

Reinforced concrete noncylinder pipe is one of the types of concrete pipe that has a limited application. Because it does not contain a watertight membrane (steel cylinder), it is used for internal pressures of 55 psi or less.

Typical applications are low pressure transmission lines used for irrigation, industrial or domestic raw-water supply and discharge lines, sanitary and storm sewers, and drainage culverts. The use of this pipe has been declining since the introduction of prestressed concrete cylinder pipe.

Size range and design basis. Reinforced noncylinder pipe is commonly furnished in diameters of 12-144 in., but larger diameters can be furnished if shipping limitations permit. Standard lengths are 8-24 ft, with the C302 specification limiting the maximum length that can be furnished for each pipe size. The design appendix in C302 covers the specific design procedure, including design for external load and internal pressure individually and in combination. As stated previously, the maximum working pressure is limited by specification to 55 psi. Earth covers up to 20 ft are common.

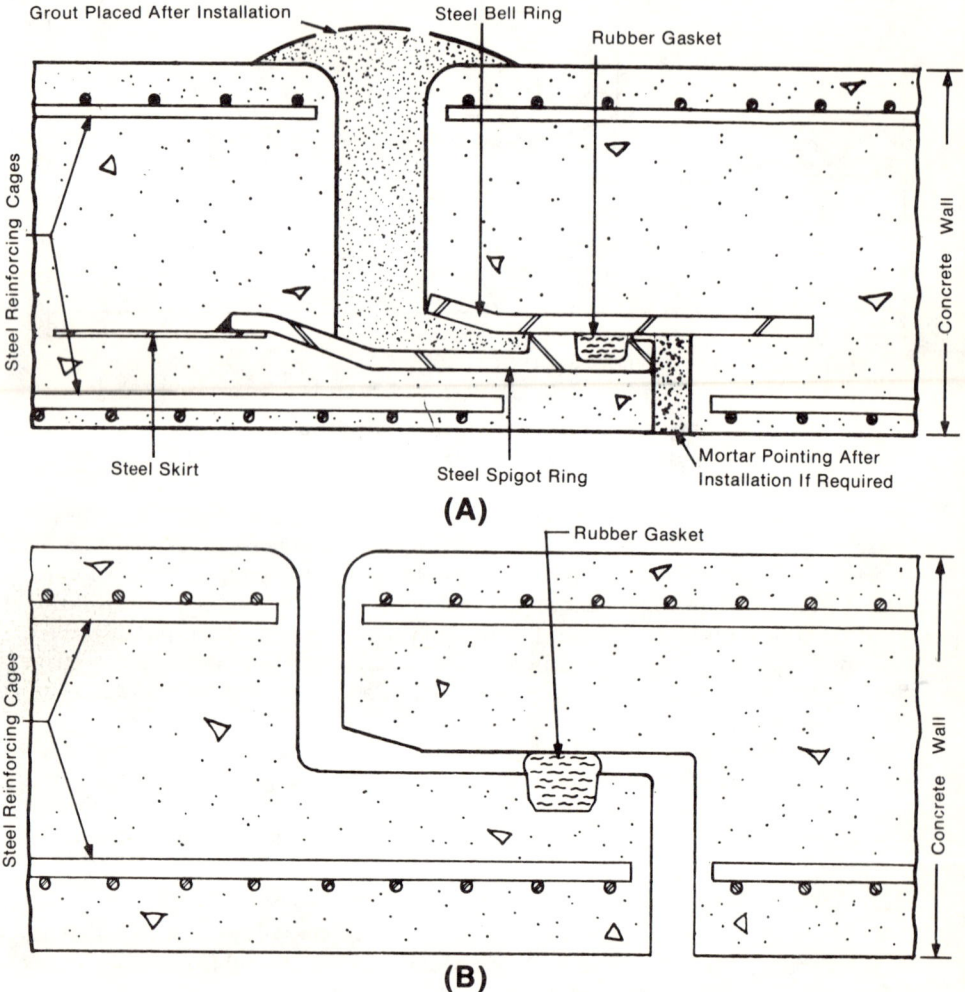

Fig. 2.3 Reinforced concrete noncylinder pipe

(A) Typical cross section using steel joint rings.
(B) Typical cross section with concrete bell and spigot joint.

Joints. The standard joint for reinforced noncylinder pipe is illustrated in Fig. 2.3. As with both reinforced and prestressed concrete cylinder pipe, this joint incorporates a continuous solid rubber ring gasket placed in a spigot groove and compressed by the flat bell surface. However, the bell and spigot ends of reinforced noncylinder pipe may be steel joint rings or formed concrete surfaces. If steel joint rings are used, the grouting procedures are the same as described previously for C301 and C300 pipe.

Manufacture. Manufacture of reinforced concrete noncylinder pipe begins with fabrication of one or more reinforcing cages (depending on pipe size and wall thickness) and fabrication of steel joint rings if that type of joint is used. When concrete is to be placed by the vertical casting method, cages and joint rings are positioned between inner and outer forms. When concrete is to be placed by the centrifugal process, cages and joint rings are secured to an outer form and the entire assembly is rotated at high speed while the concrete is placed. The concrete is cured by steam or water.

Pretensioned Concrete Cylinder Pipe (C303)

Pretensioned concrete cylinder pipe is manufactured mainly in the western and southwestern areas of the US for moderate and high pressure service. It is normally available in diameters of 10-42 in., but larger diameters can be manufactured. Typical applications are cross-country transmission mains, distribution feeder mains, sewer force mains, and plant piping. Standard lengths are generally 32-40 ft.

Manufacture. Manufacture of pretensioned concrete cylinder pipe begins with a fabricated steel cylinder with joint rings, which is hydrostatically tested. A cement-mortar lining is placed by the centrifugal process inside the cylinder. The lining is ½-in. thick for sizes up to and including 16 in., and ¾-in. thick for larger sizes. After the lining is cured by steam or water, the cylinder is wrapped with smooth hot-rolled steel bar, using a moderate tension in the bar. The size and spacing of the bar, as well

Steel cylinder with joint rings on each end being hydrostatically tested to ensure water tightness of welds

as the thickness of the steel cylinder, are proportioned to provide the required pipe strength. The cylinder and bar wrapping are covered with a cement slurry and a dense mortar coating that is rich in cement. The coating is cured by steam or water.

Joints. The standard joint detail, as shown in Fig. 2.4, consists of steel joint rings and a continuous solid rubber ring gasket. The exterior joint recess is normally grouted and the internal joint space may or may not be pointed with mortar.

Design basis. The C303 specification contains an Appendix A which covers the procedures for designing pretensioned concrete cylinder pipe. The design of this pipe is based on semirigid pipe theory in which internal pressure and external load are designed for separately but not in combination. Pretensioned pipe can be designed for internal pressures up to 400 psi. Since the theory of semirigid pipe design is based on the restraint of the soil adjacent to the springline of the pipe for resisting external load,* the design of this pipe must be closely coordinated with the installation conditions.

Prestressed Concrete Noncylinder Pipe

Prestressed concrete noncylinder pipe was first introduced in France in 1937. There have been only a few installations of prestressed noncylinder pipe in the US and Canada and, as a result of the limited use of the product, there is no AWWA standard for it.

Fittings and Special Pipe

All concrete pressure pipe manufacturers make a wide variety of fittings and special pipe, including elbows, tees, wyes, crosses, manifolds, reducers, adapters, wall sleeves, closures, bulkheads, bevel pipe, and service outlets. Adapters and

Concrete core being placed in lined cylinder pipe by the centrifugal process using trunion spinners

---

*Bulletin 153, Dec. 1941, Iowa State Experiment Station, by Merlin Spangler.

DESCRIPTION OF CONCRETE PRESSURE PIPE

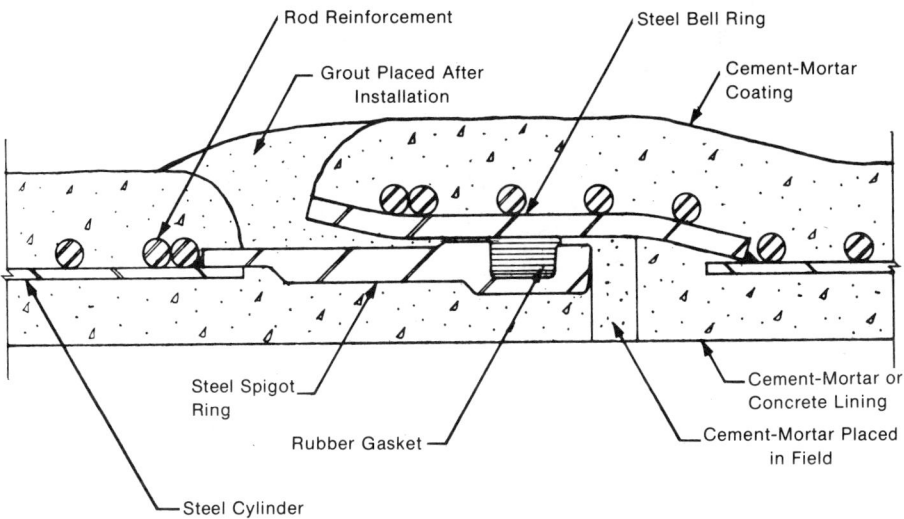

Fig. 2.4 Pretensioned concrete cylinder pipe

Cured concrete core has been helically wrapped with prestressing wire and is ready for off-bearing

outlets are available to mate with many other types of joints, such as mechanical (bell or spigot), flanged, Victaulic, threaded (standard or Mueller), plain end (for Dresser coupling or field weld to steel pipe), and cast-iron (bell or spigot for leaded joints).

Types of fittings. The AWWA standards for concrete pressure pipe (C300, C301, C302, and C303) specify the construction of fittings. For C300, C301, and C302 pipe, the standards specify either Type A or Type B fittings. Type A fittings consist of a steel cylinder lined with cement-mortar or concrete and covered with a cage-reinforced concrete coating. Type B fittings are made of cut and welded steel plate with mesh reinforcing attached to the inside and outside surfaces of the steel plate and cement-mortar applied to minimum thicknesses of ¾ in. inside and 1 in. outside. Most concrete pressure pipe manufacturers make one type of fitting (A or B) as their standard for use with C300, C301, or C302 pipe. For C303 pipe, the standard calls for fittings made of welded steel sheet or plate-lined and coated with cement-mortar. The exterior mortar coating is always wire mesh reinforced. The mortar lining may or may not be reinforced depending on the pipe size and the method used to place the lining.

Design of fittings. Fittings for concrete pressure pipelines are designed to have a

A lined cylinder type prestressed pipe being coated with mortar on a horizontal coating machine

Installation of a large-diameter wye

pressure rating equivalent to the adjacent pipe. Outlet areas are reinforced by means of saddle plates, reinforcing pads, collars, or crotch plates. Sizing of crotch plates is simplified by a series of nomographs published in the June 1955 issue of *Journal AWWA*.[1] The design of saddle plates, reinforcing pads, and collars is based on providing additional steel adjacent to an opening to resist the loads and stresses imposed.

Special pipe. Special pipe is usually defined as pipe in which basic construction is the same as the standard lengths (prestressed cylinder, reinforced cylinder or noncylinder, or pretensioned cylinder) but contains some modification such as an outlet or a beveled joint ring. Many types of outlets can be furnished more economically in a special pipe instead of using a fitting. Bevel pipe are manufactured with a joint ring deflected at a slight angle (up to 5 deg maximum). An individual bevel pipe or a series of bevel pipe can be used to deflect a line vertically or horizontally to avoid other utilities' lines, or a series of bevel pipe can be used to form a long sweeping curve to eliminate the need for an elbow and thrust restraint. Most manufacturers make bevel pipe with more than one angle of deflection, which, together with standard lengths, provide many layout possibilities for offsets and curves without the use of elbows.

References

1. SWANSON, H.S.; CHAPTON, H.J.; WILKINSON, W.J.; KING, C.L.; & NELSON, E.D. Design of Wye Branches for Steel Pipe. *Jour. AWWA,* 47:6:581 (Jun. 1955).

CHAPTER 3
Hydraulics

Flow Formulas
The following data and information is not intended to be used as a textbook, but as a guide that engineers and water utility managers can use for the design of concrete pipelines. Therefore, it is limited to a range of velocities and diameters associated with concrete pressure pipe for water transmission. Practical application of the flow formulas in this chapter have been prefaced by a general history of the development of empirical and rational formulas.

Empirical Formulas
Many empirical flow formulas are found in the technical literature. Each of these expresses some relationship between flow, hydraulic gradient, pipe size, and resistance coefficient. A comparison of the formulas shows that they differ in the form of resistance coefficient used, and in the power to which the key variables of velocity, diameter, and hydraulic gradient are raised. In effect, they are basically different statements of flow behavior. These differences are manifested mostly in the way head loss is expressed as a function of velocity.

Chezy.   The basis of all flow formulas is the expression

$$V = C_c r^{0.5} S^{0.5}$$

in which V = velocity, in feet per second; C_c = the Chezy roughness coefficient; r = hydraulic radius, in feet; and S = slope of the hydraulic gradient, in feet per foot.

Darcy-Weisbach.   This expression was introduced by Chezy in 1775 for flow in open channels. Darcy and Weisbach later rewrote the equation for use in round pipe, replacing the Chezy coefficient with the radical $\sqrt{8g/f}$, thus making the Chezy formula dimensionally correct. The Darcy-Weisbach expression is

$$h_f = f \frac{LV^2}{2Dg}$$

in which h_f = friction head loss for the test length, in feet; f = Darcy friction factor; L = length, in feet; V = velocity, in feet per second; D = internal diameter of the pipe, in feet; and g = acceleration due to gravity, in feet per second per second.

The basic limitation of this expression was the value of the friction factor f. Tests showed that this factor depended on the velocity, pipe diameter, pipe wall surface and, to some extent, the density and viscosity of the fluid. It may be noted that if the factor f is itself a function of velocity, then head loss becomes a function of some power of the velocity other than 2.

Further modifications of Chezy's expression.   Kutter and Manning attempted to reformulate Chezy's expression so that the coefficient would remain constant for a given wall surface. Many such empirical expressions have been proposed, but the Hazen-Williams and Scobey formulas are most often used in the western hemisphere.

Hazen-Williams formula.[1] Published in 1905, the Hazen-Williams formula is

$$V = 1.318\, C_h r^{0.63} S^{0.54}$$

in which C_h = the Hazen-Williams roughness coefficient.

This formula may be expressed in terms of head loss:

$$h_f = \left(\frac{1.816}{C_h}\right)^{1.852} \left(\frac{L}{D^{1.167}}\right) V^{1.852}$$

The Hazen-Williams formula presented changes both in the exponents of pipe diameter and velocity. It was based on test data from several types of pipe that had relatively smooth surfaces.

Scobey's formula. Extensive studies of test data for the US Department of Agriculture[2-4] led Scobey to the important conclusion that pipe of different wall surface characteristics resists flow according to different laws. Thus, he developed separate flow formulas for riveted-steel, wood-stave, and concrete pipe. The formula for concrete pipe in the terms used in this manual is

$$V = 355\, C_s r^{0.625} S^{0.5}$$

in which C_s = the Scobey roughness coefficient; r = the hydraulic radius; and S = the slope, in feet per foot.

In terms of head loss, this formula is

$$h_f = \left(\frac{0.00669}{C_s}\right)^2 \left(\frac{L}{D^{1.250}}\right) V^2$$

in which D = the diameter, in feet; and V = the velocity, in feet per second.

This formula was presented in 1920. The data for its basis came from concrete pipelines built in 1883. Because the concrete pipe built then was substantially rougher than the pipe on which the Hazen-Williams formula is based, the head loss computed by the Scobey formula reverted to a function of the square of the velocity as it had been formulated by Manning and Chezy years before.

Changes in actual pipe roughness. Prior to the 1920's, pipes were often built with wooden forms, and in short lengths. Joints were not formed to close tolerances, and little care was taken in alignment or joint finishing. This contrasts with present-day manufacture, which employs steel forms, centrifugal spinning, high-speed mechanical vibration, carefully machined joint-forming rings, self-centering gasket joints, and longer lengths. It is significant to note that present-day concrete pipe surfaces are similar to those on which the early Hazen-Williams formula was based and confirmed. This would suggest that the variation in head loss would more closely follow Hazen-Williams than Scobey, which would be in agreement with Scobey's own findings: that different pipe resist flow according to different formulas.

Mills' formula. Many other empirical formulas have been advanced and found to be useful for practical design. One of the earliest, by Mills, has been strongly advocated by Capen.[5] The formula (in the notation of this manual) is

$$V = 75.2\, F_m r^{0.625} S^{0.5}$$

in which F_m = the Mills roughness coefficient. It can be seen that Scobey's formula for concrete pipe is quite similar.

Rational Flow Formulas

Although engineers who are more interested in the immediate and practical considerations of design have developed purely empirical formulas, such as those just previously discussed, other investigators have attempted to establish flow relationships on a more rational basis. These attempts have taken the form of evaluation of f in the dimensionally correct Darcy-Weisbach equation.

Hagen, Poiseuville, and Reynolds. The earliest of this work was done in the 19th century by Hagen, Poiseuville, and Reynolds, who formulated expressions that related the effect of fluid density and viscosity to frictional losses. Hagen and Poiseuville developed an exact equation for laminar flow, and Reynolds defined the limits of laminar and turbulent flows.

Later developments: Von Karman to Moody. In the 1930's Von Karman, Prandtl, and Nikuradse conducted theoretical and experimental studies that led to fairly precise formulas for two types of turbulent flow. Based on the laws of mechanics, the formulas expressed the relationship of the Darcy friction factor f to the Reynolds number and to the relative roughness of the pipe wall. The two types of flow are characterized as (1) flow in smooth pipe (such as glass, brass, and copper), and (2) fully turbulent flow in rough pipe. Later, Colebrook and White[6] developed a transition equation which bridged the gap between flow in smooth pipe and that in the fully turbulent range. This transition, meant to be applicable to most types of commercial pipe, was empirically derived from tests on new wrought-iron, cast-iron, and tar-coated iron pipe. All this work culminated in a comprehensive diagram presented by Moody,[7] which reduced the complicated equations for flow to a relatively simple diagram that plots the Darcy friction factor against the Reynolds number for various values of relative roughness.

Application of the Moody diagram: Rouse. The proper use of the Moody diagram rested on the establishment of dependable values of the roughness parameter e. Rouse[8] proposed values for various types of commercial pipe. Those for concrete pipe were apparently derived by assuming the equal validity of the Manning and Von Karman rough-pipe formulas in the fully turbulent zone. This is expressed as

$$0.093 \frac{r^{1/6}}{n} = 2 \text{ Log} \frac{r}{e} + 2.35$$

in which n = the Manning coefficient; and e = the Moody absolute roughness, in feet.

Relationship of Manning's n *to Moody's* e. Using representative values of the Manning n, the roughness parameter e was calculated. Inspection of this equation, however, shows that no single-value relationship exists between n and e because the dimension of the pipe must be included. Moreover, because the flow in concrete pressure pipe is not in the fully turbulent range, the relationship yields values of e that are far too conservative for concrete pressure pipe. These values are shown along with the Moody diagram in many tests on hydraulics, and the matter has been mentioned to correct a seriously misleading point.

Types of Flow

Basic to the construction of the Moody diagram is the hypothesis that the flow of fluids through conduits exhibits three distinct variations in physical behavior.

These variations occur in different ranges of velocity, and are called *laminar, transition*, and *fully tubulent flow*.

Transition from laminar to turbulent flow. In laminar flow the roughness of the pipe wall does not affect the variation in head loss as velocity increases because the distribution of velocity across the pipe is parabolic and is zero at the wall surface. Laminar flow, which occurs over a limited range of velocity (well below those common to water transmission), is characterized by smoothly flowing streamlines. When velocity is increased beyond this range, friction resistance along the wall causes a disturbance in the streamline. A turbulence is thereby initiated, which is reflected across the pipe, except for a thin boundary layer of laminar film along the wall into which it does not extend. If the wall is perfectly smooth, the resistance will be a function of the viscous properties of the film. If the wall is rough, then roughness elements will project up through the film and contribute to resistance.

The amount of resistance contributed depends on the thickness of the laminar layer, which thins out as velocity of flow increases. At sufficiently high velocities the film is theoretically entirely dissipated, and resistance is a function of roughness elements only. The "fully turbulent" resistance then remains unchanged even with further increases in velocity. At lower velocities, however, the laminar film will only partially cover the roughness elements, making resistance a complete function of film and roughness. This function will constantly change as increasing velocity thins the film, thus providing a gradual transition between essentially smooth pipe action and fully turbulent flow.

Turbulent flow. More recently, Morris[9,10] has published design diagrams derived from the basic mechanics of flow for different types of turbulence created by distinctly different types of pipe roughness. The types of flow covered are semi-smooth flow over isolated roughness elements, skimming flow over various kinds of grooves, and hyperturbulent flow generated by extreme roughness such as corrugations or sand grains. Each type of roughness exhibits a distinctive resistance, requiring its own formula or diagram. This work appears to contradict, in rather fundamental ways, the basic assumptions manifested in the Moody diagram. Not only is the idea of a universal diagram that is applicable to all types of pipe questionable, but the patterns of flow in the several diagrams differ markedly. Of particular interest is the formulation of isolated-roughness flow, which discards the idea of fully turbulent flow and retains a laminar boundary film between the isolated roughness elements. The result is that a family of curves, depending on the height, shape, and frequency of the projections, is produced with the Von Karman smooth-pipe curve at its lower limit. This theory provides justification for a decreasing friction factor beyond velocities comprehended by the Moody diagram. This fact has been made evident by experimental data and has found empirical expression in the Hazen-Williams formula.

Comparison of Formulas

The problem of finding dependable flow formulas has been studied for years. True progress is never achieved until the fundamental laws governing a physical phenomenon are understood and expressed mathematically. To this end, investigation into principles inherent in turbulent flow is still being actively pursued, but differences of opinion on final conclusions still remain. Exception has also been taken, from time to time, to the application of empirical formulas. Much of this criticism has been based on the assumed validity of the theory of transitional flow (Colebrook-White equation) shown in the Moody diagram. The compatibility of the theoretical isolated-roughness formula of Morris and the empirical Hazen-Williams expression suggests that the criticism was unjustified to the extent that this particular empirical formula is applicable to pipe of a certain type. Morris's other types of flow present more cases that differ considerably from that described by the Moody diagram. This is not meant to suggest that the Moody diagram is not a useful tool, but only that it is chiefly applicable to pipe of certain surface characteristics.

Problems with correlation. Since it is still not possible to say that one formula is superior to all others for *all* types of pipe, it is instructive to see how they are related to one another. This comparison may best be shown by the way in which the Darcy friction factor f changes with the Reynolds number N_r, as illustrated in Fig. 3.1. In the figure, the Darcy friction factor f is defined as

$$f = \frac{h_f}{\frac{LV^2}{2Dg}}$$

and the Reynolds number N_r can be expressed as

$$N_r = \frac{VD}{\gamma}$$

in which γ = kinematic viscosity, in square feet per second.

The Colebrook-White equation shows that f decreases as N_r increases up to the point termed fully turbulent flow. Beyond this point f remains constant. The Hazen-Williams equation may be rewritten in terms of f and N_r, and in this form it reflects flow below the so-called fully turbulent range, as a constantly decreasing f results from an increasing N_r. The Scobey formula only reflects flow of the fully turbulent variety because a constant f is indicated. This reflects original test conditions, because the tests that resulted in the formulation of the Hazen-Williams equation were performed on smooth pipe, and those that resulted in the Scobey equation were performed on rough pipe. Morris's isolated roughness theory shows that f decreases with increasing N_r, but never becomes a constant, thus strongly supporting in theory the variation in f as previously determined by Hazen and Williams. In passing, it might be noted that the Chezy, Manning, and Mills formulas take the same form as the Scobey equation in this diagram.

General results. Inspection of the curves in Fig. 3.1 shows a wide divergence for high and low velocities, but rather good correlation at a velocity of 4-5 fps. Because of the economics involved in determining the optimum size associated with minimum capitalized cost, the velocities in water pipelines usually fall within the range of 2-10 fps. Diameters of commercial concrete pipe for pressure service generally range from 14 to 144 in. in diameter. The problem is rather limited to a narrow range of Reynolds numbers.

Practical Application

A detailed investigation of the available flow test data for concrete pipe was made by Swanson and Reed.[11] Their conclusion was that, for the range of velocities normally encountered in water transmission, the empirical Hazen-Williams

CONCRETE PRESSURE PIPE

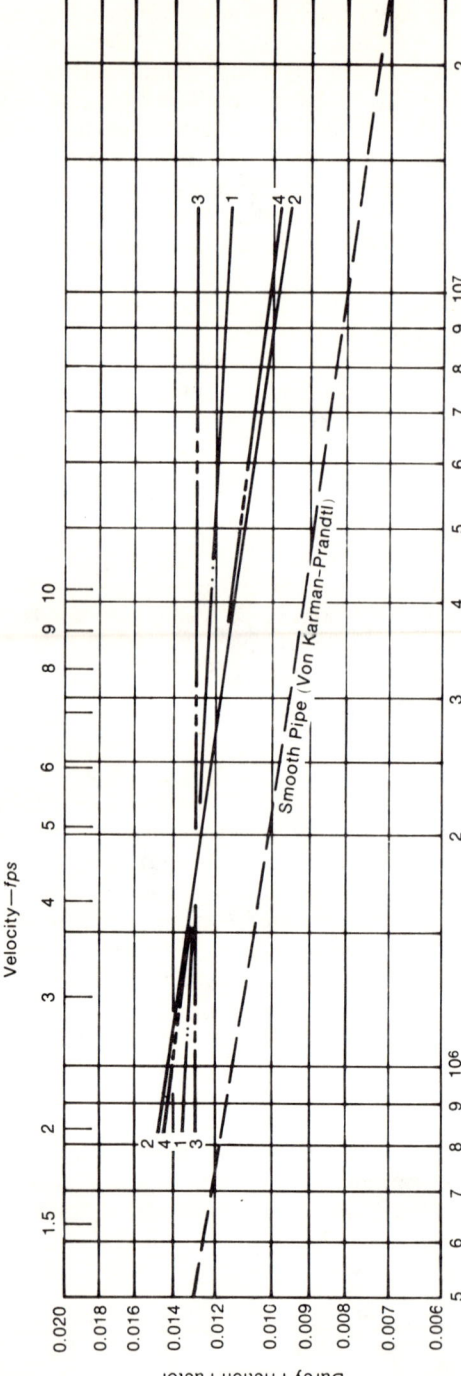

Fig. 3.1 Comparative study of frictional resistance for 60-in. cast concrete pipe

Curves are for water temperature of 60°F (16°C) and design coefficients as recommended in the text. Curve 1 represents the Colebrook-White transitional formula of the Moody diagram; Curve 2, Morris isolated roughness; Curve 3, Scobey for concrete pipe; and Curve 4, Hazen-Williams.

HYDRAULICS

formula most closely matches the test results. This formula has the advantages of easy application and acceptable accuracy.

Swanson and Reed analysis. The equation in Figure 3.2 was developed from a statistical analysis of 67 tests reported by Swanson and Reed, with the exception that data defined as "cast concrete tunnels" was deleted. The method employed consisted of a regression analysis least-squares method, which produced the correlation equation

$$C = 139.3 + 0.169D$$

In Table 3.1 the actual value of each of the 67 tests is compared to its theoretical value from the equation as a percent of theoretical.

Suggested Hazen-Wiliams design values. Although it is possible to calculate a different C value for each pipe size, the following conservative values are suggested for design:

Diameter—in.	C Value
16-48	140
54-108	145
114 and larger	150

These values are applicable to concrete pipelines where the fitting losses are a minor part of the total loss. These values also assume that the line is free from organic growths or chemical deposits that can materially affect the pipe's carrying capacity.

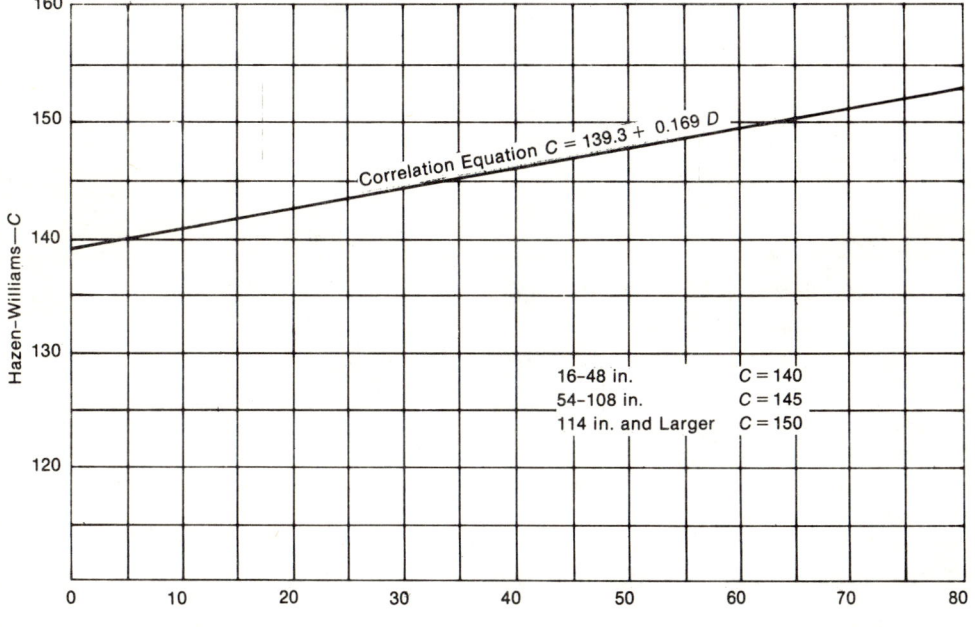

Fig. 3.2 Hazen-Williams coefficient

TABLE 3.1
Comparison of Theoretical Hazen–Williams C Values to Tested C Values

Sample Number	Diameter	C Measured	C Theoretical	Percent of Theoretical
1	24.0	145.0	143.4	101.15
2	30.0	142.0	144.4	101.82
3	30.0	145.0	144.4	100.44
4	31.5	147.0	144.5	101.64
5	31.8	152.0	144.7	105.06
6	36.0	143.0	145.4	98.36
7	36.0	150.0	145.4	103.17
8	36.0	141.5	145.4	97.33
9	36.0	142.5	145.4	98.02
10	36.0	150.0	145.5	103.17
11	39.0	147.0	145.9	100.76
12	42.0	147.5	146.4	100.75
13	42.0	142.0	146.4	97.00
14	42.0	149.0	146.4	101.78
15	42.0	149.0	146.4	101.78
16	42.0	142.0	146.4	97.00
17	42.0	147.0	146.4	100.41
18	46.0	144.0	147.1	97.91
19	48.0	146.0	147.4	99.04
20	48.0	150.0	147.4	101.75
21	48.0	144.0	147.4	97.69
22	51.0	142.0	147.9	96.00
23	54.0	152.0	148.4	102.41
24	54.0	141.0	148.4	95.00
25	54.0	140.5	148.4	94.66
26	60.0	145.5	149.4	97.36
27	60.0	152.5	149.4	102.05
28	46.0	148.5	147.1	100.97
29	54.0	150.0	148.4	101.06
30	36.0	138.0	145.4	94.92
31	36.0	137.0	145.4	94.23
32	36.0	142.0	145.4	97.67

An alternate solution. An alternate solution to the Hazen-Williams equation is the Darcy-Weisbach expression

$$h = f \frac{LV^2}{2Dg}$$

Values of f can be determined from Fig. 3.3, "The Moody Diagram." The Moody e values determined from the data used for the C factor curve ranged from 8×10^{-5} ft to 7.5×10^{-4} with an average value of 2×10^{-4}. Conservative values of $3.5-4 \times 10^{-4}$ are recommended for design purposes. This range will give good correlation with Hazen-Williams results in 3-7 fps velocity range as indicated in Fig. 3.1.

Influence of Large Reynolds Numbers

It should be noted that the ASCE Task Force Report [12] states that, "A careful study of field tests by the United States Army Engineers Waterways Experimental Station has resulted in the conclusion that Eq. 9 (Colebrook-White equation)

TABLE 3.1 (cont.)

Sample Number	Diameter	C		
		Measured	Theoretical	Percent of Theoretical
33	48.0	149.0	147.4	101.08
34	54.0	151.5	148.4	102.07
35	54.0	145.0	148.4	97.69
36	54.0	146.0	148.4	98.37
37	54.0	147.5	148.4	99.38
38	60.0	147.0	149.4	98.37
39	60.0	154.0	149.4	103.05
40	60.0	156.0	149.4	104.39
41	60.0	143.5	149.4	96.03
42	66.0	142.0	150.5	94.38
43	48.0	156.5	147.4	106.17
44	48.0	152.0	147.4	103.11
45	48.0	149.5	147.4	101.42
46	48.0	153.5	147.4	104.13
47	54.0	154.0	148.4	103.76
48	54.0	155.5	148.4	104.77
49	54.0	151.0	148.4	101.73
50	54.0	152.0	148.4	102.41
51	54.0	143.0	148.4	96.34
52	72.0	154.0	151.5	101.67
53	12.0	145.0	141.3	102.60
54	18.0	147.0	142.3	103.27
55	20.0	134.0	142.7	93.92
56	24.0	149.5	143.4	100.10
57	30.0	141.0	144.4	97.67
58	30.0	143.0	144.4	99.05
59	33.0	147.0	144.9	101.47
60	33.0	146.0	144.9	100.78
61	33.0	146.0	144.9	100.78
62	36.0	134.0	145.4	92.17
63	31.4	138.0	144.6	95.43
64	36.0	147.0	145.4	101.11
65	36.0	150.0	145.4	103.17
66	41.9	150.0	146.4	102.47
67	42.0	150.0	146.4	102.46

cannot be verified for large Reynolds numbers." This same reference defines large Reynolds numbers as being greater than 10^6.

Effects of Aging on Carrying Capacity

Because concrete does not corrode, the carrying capacity is not affected by age. There are, however, several organic growths and some types of inorganic chemical deposits that can produce a reduction in carrying capacity and develop layers of substantial thickness. These conditions can occur on any type of piping material and are generally associated with raw-water transmission mains. There are means of removing these growths and deposits from concrete lines, thereby restoring the line to its original capacity.

Minor Losses

Flow-through fittings, valves, and appurtenances produce head losses that add to

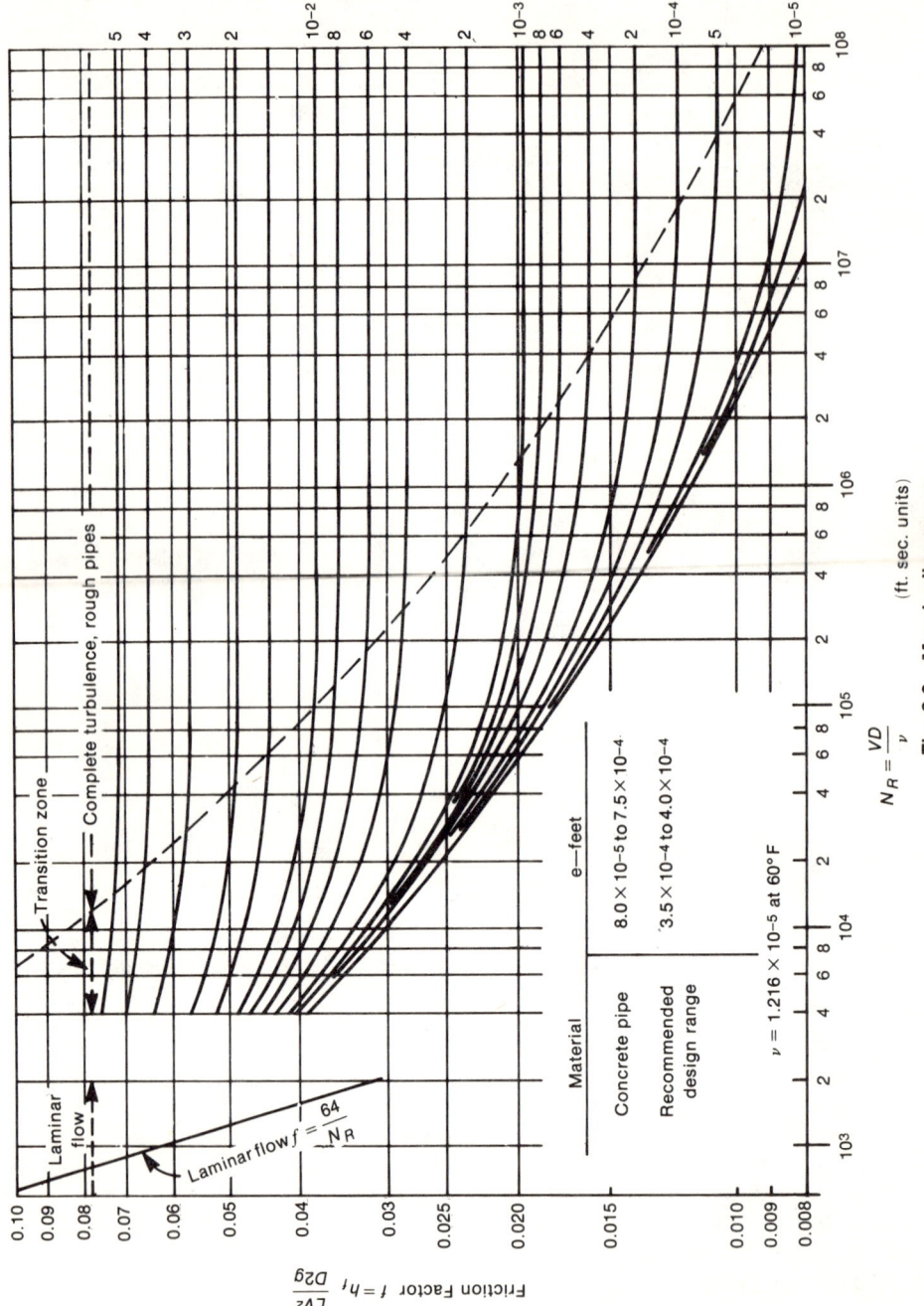

Fig. 3.3 Moody diagram

friction losses. In pipe sizes normally associated with water transmission and distribution these losses are generally referred to as minor. However, they can be an important factor in the design of treatment plants and similar types of systems.

The rational method. The rational method assumes full turbulence at those points and expresses the loss in terms of velocity head

$$h = K \frac{V^2}{2g}$$

where K = a dimensionless coefficient, and V = mean velocity.

There are several published studies and tests that give suggested values for K.

Commonly used K *values.* The values for K in Table 3.2 have been commonly used for design purposes.

TABLE 3.2
Fitting Losses

Fitting	K
Gate valve	
Fully open	0.2
¼ closed	1.2
½ closed	5.6
¾ closed	24.0
Swing check valve	2.0
Elbows	
90 deg	0.6
45 deg	0.5
Tee, through run	1.8
Sudden contraction	
$d/D = ¼$	0.4
$= ½$	0.3
$= ¾$	0.2
Sudden enlargement	
$d/D = ¼$	0.9
$= ½$	0.6
$= ¾$	0.2
Flow into storage or reservoir	1.0
Entrance	0.5

Significance of size. It should be noted that many concrete pipelines are made in the 30-in. and larger diameters and that some of the tests were based on small diameter pipe, 12 in. and smaller, and may not be applicable. The designer should be careful to select values that are consistent with the pipe diameter and the kind of fittings in question. For large diameter pipe the best information available is contained in the ASCE Task Force Report. This report incorporates comparative data that a designer might use to advantage on a specific project relative to entrance losses, trash racks, gate valves, butterfly valves, elbows, and wyes.

Equivalent length method. For the degree of accuracy required for most projects the loss of head in a fitting or special pipe can be expressed by the equivalent length method. The equivalent length is the number of feet of straight pipe having the same loss as the fitting or appurtenance. This length is added to the total length of the pipeline involved and losses are calculated as friction losses. Equivalent lengths for various fittings and specials are in the Crane charts.[13,14] A method of converting K values to equivalent lengths has been published by Labella[15] for appurtenances not shown in the Crane chart.

Valves, water meters, and other specials that create variable losses because of shape differences or use characteristics should be considered as special cases and the flow-discharge curves should be obtained from the manufacturers for accurate determination of losses.

Air Entrapment and Release

Air binding. Air entrapment in poorly vented pipelines can result in air binding with an accompanying loss of head and reduction of flow. Three conditions must exist before air binding presents a serious problem:
1. There must be a source of air in excess of that normally held in solution in the flow.
2. The air must separate from the water at high points.
3. The collected air must remain in equilibrium keeping it from moving along the conduit.

The sources of the air can be any one of several. The most common sources are air entrapped when filling, entrainment at the intake by vortex action, and hydraulic jump as a result of partial gate opening. It is desirable that the high points on transmission mains be properly vented to relieve this condition. One consideration overlooked in many designs is the air and vacuum valve itself. If the valves are too small, the air velocities can become extreme increasing the danger of damage to the valve.

Water Hammer

General. The fundamental relationships associated with water hammer are flow velocity, pressure wave velocity, and rate of flow cutoff. These were established by Jonkowsky in 1897. The principles have been confirmed and expanded into present-day analysis procedures, some of which are very complex and require a good deal of specialized knowledge.

Standard allowance. Concrete pressure pipe made in accordance with AWWA C301 has built-in allowance equal to at least 40 percent of the working pressure in combination with earth loads. Pipe made in accordance with AWWA C303 has an allowance of 50 percent of the working pressure. Pipe made in accordance with AWWA C302 incorporates the water hammer requirements in the design pressure. In the latter case the pressures involved are 65 psi or less and water hammer is seldom a problem. In most design problems this allowance will provide the necessary margin. There are, however, line configurations and pressure ranges that should be investigated for water hammer potential. Examination of the basic relationships should be helpful in deciding when a particular solution can be used safely.

Basic parameters. The basic or fundamental parameters that must be established for water hammer calculations are the following:
1. Flow velocity V, which can be calculated easily from pipe diameter and flow

rate. For many years 4-5 fps was considered a maximum value in water works practice; this has gradually increased to 10 fps, particularly in transmission mains.
2. Pressure wave velocity a, which is expressed by the formula

$$a = \frac{4720}{\sqrt{1+\frac{KD}{Et}}}$$

where K = bulk modulus of water = 300,000 psi; D = internal diameter, in inches; E = modulus of elasticity of the pipe wall; and t = wall or core thickness. Since concrete pipe is a composite structure, the apparent modulus of elasticity varies with the strain induced in the pipe wall. This value will vary from the condition where the concrete supports the tensile loads at one extreme to the condition where the steel components support the load at the other extreme. A detailed explanation of the theory has been published by Kennison[16] together with a range of values for various kinds of concrete pipe. Design values for types of concrete pipe are listed in Table 3.3.
3. Length of the pipeline L, where length is defined as a single line of uniform diameter, material, and wall thickness. If these quantities vary in a single line, the line is treated in sections; each being analyzed separately. The presence of a single connection can complicate the analysis or greatly reduce the reliability of the results.
4. Time effect T defined as the closing time, in seconds, of the control device, assuming that the velocity increases or decreases at a constant rate. Since valves characteristically do not produce a uniform rate of change due to their shape, the effective time T_e must be established for each kind of valve. This is the time in which the valve must be open or close to give a uniform rate of velocity change equal to the established maximum rate. T_e varies from 20 to 50 percent of actual time of valve closing.

Maximum pressure rise. The maximum pressure rise for instantaneous flow stoppage is

$$h = 2.31\, p = \frac{aV}{g}$$

in which h = feet; p = pounds per square inch; a = feet per second; V = feet per second; and g = feet per second per second.

If the value of a = 3220 or 100 × g, then the maximum water hammer pressure rise

TABLE 3.3
Design Values - Surge Wave Velocities - fps

AWWA Description		Pipe Class – *psi*		
		100	150	200
Prestresed concrete pressure pipe, steel cylinder type (C301)	Lined cyl.	3500-3750	3400-3775	3350-3850
	Embedded cyl.	3025-3625	3000-3650	2975-3675
Reinforced concrete water pipe, steel cylinder type, pre-tensioned (C303)		3175-3875	3325-3925	3450-3975

will be 100 ft per foot of velocity head change. A conservative rule of thumb measurement is 50 psi per foot of velocity head change.

In the case of incomplete flow stoppage, the maximum pressure rise is

$$h = 2.31\, p = \frac{a}{g}(V_1 - V_2)$$

in which V_1 = initial velocity, and V_2 = final velocity.

Critical period. The critical period of a pipeline is the time required for a pressure wave to travel from the point of interruption to the point of origin of flow and return. It is expressed as

$$t_c = 2L/a$$

In the simplified case of a gravity line with lower end control operating under a positive head, a straightforward analysis is possible. When the closure is instantaneous the magnitude of the surge remains constant for the full length of the line. When the closure time is greater than zero but less than $2L/a$, the surge travels a distance equal to $L - t_c a/2$ and then decreases uniformly. When the closure time is equal to $2L/a$ the maximum pressure occurs at the point of closure and decreases uniformly to zero at the point of origin of flow. For a closure time greater than $2L/a$, the maximum pressure will be a function of the rate of change of V with regard to time. The greatest incremental pressure change will result from the largest velocity change that occurs during the critical period of the line $2L/a$.

If it is assumed that $t_c = 1.5$s (which is a period of time brief enough so that a significant change in velocity cannot occur in the pipe sizes that are usually encountered with concrete pipe) and a is assumed to be 3333 fps, then the length of the line for this critical period would have to be 2500 ft. Under these conditions maximum water hammer is not likely to occur, the single exception being lines equipped with swing type check valves, which should be avoided.

Other considerations. The information presented so far in this manual will allow the designer to analyze the simpler water hammer problems and to exercise judgment in determining whether approximate solutions are appropriate.

It must be recognized, however, that some water hammer problems are complex and will require specialized knowledge in order to attain a reasonable and economical solution. The designer should satisfy himself that he is sufficiently knowledgeable regarding water hammer theory before he attempts to solve the more complex problems.

Control of water hammer in transmission mains is more economical than designing the entire pipe system to withstand excessive surge pressure.

References

1. WILLIAMS, G.S. & HAZEN, A., *Hydraulic Tables.* John Wiley & Sons, New York, (3rd ed., revised 1933).
2. SCOBEY, F.C. The Flow of Water in Concrete Pipe. Bul. No. 852, US Dept. of Agriculture, Washington, D.C. (1920).
3. SCOBEY, F.C. The Flow of Water in Wood-Stave Pipe. Bul. No. 376, US Dept. of Agriculture, Washington, D.C. (1916).
4. SCOBEY, F.C. The Flow of Water in Riveted Steel and Analogous Pipes. Tech. Bul. No. 150, US Dept. of Agriculture, Washington, D.C. (1930).

5. CAPEN, C.H. Trends in Coefficients of Large Pressure Pipes. *Jour. AWWA,* 33:1 (Jan., 1941).
6. COLEBROOK, C.F. Turbulent Flow in Pipe, With Particular Reference to the Transition Region Between the Smooth and Rough Pipe Laws. *Inst. Civ. Engrs.,* Vol. 11, Paper No. 5204 (1938-39).
7. MOODY, L.F. Friction Factors for Pipe Flow. *Trans. ASME,* 66:671 (1944).
8. *Engineering Hydraulics.* Edited by H. Rouse. John Wiley & Sons, New York (1950).
9. MORRIS, H.M. Flow in Rough Conduits. *Trans. ASCE,* 120:373 (1955).
10. MORRIS, H.M. Design Methods for Flow in Rough Conduits. *Trans. ASCE,* 126:454, I (1961).
11. SWANSON, H.V. & REED, M.S. Comparison of Flow Factors for Concrete Pipe. *Jour. AWWA,* 55:1 (Jan. 1963).
12. Factors Influencing Flow in Large Conduits. Jour. of the Hydraulics Division — Proc. American Society of Civil Engineers, Report of the Task Force on Flow in Large Conduits (Nov. 1955), Discussion (Jul. 1966), Closure (May 1967).
13. Flow of Fluids Through Valves, Fittings and Pipe. Crane Company, Cat. No. 52 (1942).
14. Flow of Fluids. Tech. Paper No. 410, Crane Company (1957).
15. LABELLA, S. Equivalent Length of Pipe for Fittings, Water and Sewer Works. Ref. No. 1967.
16. KENNISON, H.G. Surge-Wave Velocity — Concrete Pressure Pipe. Paper presented Nov. 1955, *ASME,* Chicago.
17. PARMAKIAN, John. *Water Hammer Analysis.* Dover Publications (1963).

CHAPTER 4
External Loads

Introduction
Many people have contributed to the development of associated load and supporting strength theories, but the basic load theory is rightly recognized as the Marston Theory. The following presentation is based primarily upon the results of studies made by the Engineering Experiment Station of Iowa State University; many of them in cooperation with the Public Roads Administration.

Rational Approach
An outstanding contribution of the "Marston Theory of Loads on Underground Conduits" is that it demonstrates by rational principles of mechanics that the load on a structure of this kind is greatly affected by certain environmental conditions of installation as well as by the weight of fill over the conduit. The traditional view that the weight of fill is the major variable which influences the load that a conduit must be designed to carry is now augmented by several other factors which wield an important influence on the load. These factors are associated mostly with those installation conditions that control the magnitude and direction of settlements of the prism of soil directly over the conduit relative to the settlements of the exterior prisms immediately adjacent. These relative settlements generate friction or shearing forces which are added to or subtracted from the weight of the central prism to produce the resultant load on the conduit.

Major Installation Classifications
Because of the influence of these installation conditions and the importance of recognizing them when determining loads, underground conduits have been classified into groups and subgroups, as illustrated in Fig. 4.1. The two major

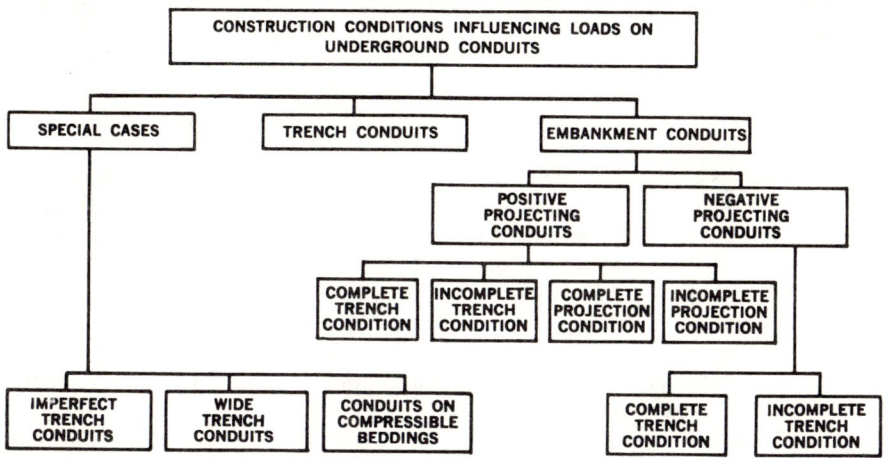

Fig. 4.1 Underground conduit classifications

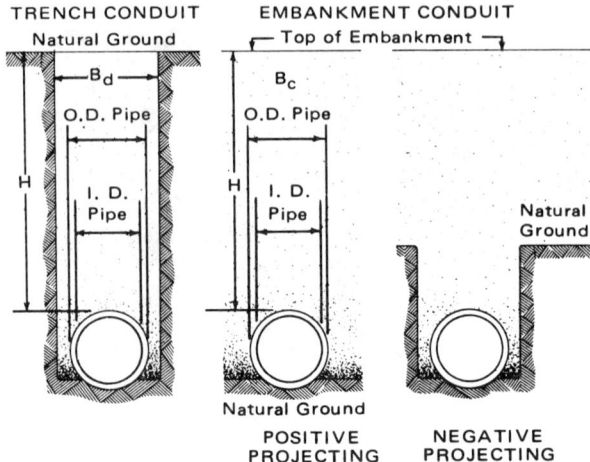

Fig. 4.2 Underground conduits

groups are trench conduits and embankment conduits. Embankment conduits are further subdivided into positive and negative projecting subgroups, depending on whether the top of the conduit, as installed, is above or below the adjacent ground surface. Essential features of these three classes of conduits are illustrated in Fig. 4.2.

Trench Conduits

The determination of the load on a conduit installed in a trench is comparatively simple. Trench conduits are installed in relatively narrow trenches dug in passive or undisturbed soil and then covered with earth backfill to the original ground surface.

Basic assumptions. Practically every theory in applied mechanics is based on a set of assumptions concerning the mechanical properties of the materials involved. In this case the assumptions are

1. Load on the conduit develops as the backfill settles because the backfill is not compacted to the same density as the surrounding earth.
2. The resultant load on an underground structure is equal to the weight of the material above the top of the conduit minus the shearing or friction forces on the sides of the trench. These shearing forces are computed in accordance with Rankine's theory.
3. Cohesion is assumed to be negligible because (a) considerable time must elapse before effective cohesion between the backfill material and the sides of the trench can develop, and (b) the assumption of no cohesion yields the maximum probable load on the conduit.
4. In the case of rigid pipe, the side fills may be relatively compressible and the pipe itself will carry practically all the load developed over the entire width of the trench.

When the conduit is placed in a trench that is not wider than two or three times its outside breadth, the backfill will tend to settle downward. This tendency for downward movement is retarded by frictional forces that develop along the sides of the trench. These forces act upward and help support the backfill.

Load formula. The magnitudes of the frictional forces are dependent upon the weight of the fill material, the value of Rankine's factor K, and the coefficient of sliding friction μ'. Rankine's factor K is a function of μ, the coefficient of internal friction of the soil, and ranges in value from 0.33 to 0.37. The fill load on the conduit is equal to the weight of the mass of fill material less the summation of the frictional load transfers, and is expressed by the formula

$$W_d = C_d\, w\, B_d$$

Where W_d = fill load, in pounds per linear foot of conduit; C_d = load-calculation coefficient; w = unit weight of fill material, in pounds per cubic foot; and B_d = width of trench at or slightly below the level of the top of the conduit, in feet.

Selection of coefficients. The values of C_d, for several soils, have been tabulated against values of the ratio H/B_d, as shown in Table 4.1. The values of $K\mu'$ and w for each case should be those that result in the maximum calculated load that is probable in each case. In some cases it may be advisable to determine the actual values of w and $K\mu'$. When the character of the soil is uncertain, it is generally adequate to assume that $w = 120$ and $K\mu' = 0.150$.

Significance of trench width. Study of the load formula and the C_d diagram in Table 4.1 shows the marked effect that the width factor B_d has on the load. An increase in the width of the trench causes a reduction in H/B_d, and also a small reduction in C_d, but since the load varies with B_d^2, an increase in B_d will cause a marked increase in load. Consequently, the value of B_d should be held to the minimum that is consistent with efficient construction operations.

Transition trench width. There is a trench width condition where the upward frictional forces are no longer effective in reducing the load on the pipe and, hence, the installation assumes the same properties as a positive projecting embankment condition. This represents the severest load that the pipe can be subjected to; any further increase in trench width would have no effect. The maximum effective trench width, where transition to a positive projecting embankment condition occurs, is referred to as the "Transition Trench Width." The load generated (that is, positive projecting) is referred to as the "Transition Load."

NOTE: The width of the trench is taken as the horizontal width at the top of the conduit. If the trench has sloping sides, the load on the pipe is equal to that for a vertical-sided trench with a width equal to the width at the level of or slightly below the top of the pipe.

Calculated loads. The preceding trench conduit formula, with proper selection of the physical factors involved, gives the maximum loads to which a particular conduit may be subjected when in service. Because of cohesion or other causes, these loads may never fully develop during the service life of the conduit.

Tables 4.2, 4.2.a, and 4.2.b are useful in determining the loads on trench conduits for some of the more common soil conditions.

Trenches in caving soils. Some projects involve the decision as to whether the trench is to be sheeted or whether the sides are to be allowed to cave in and thus produce a trench with sloping sides. When sheeting is used, the load may be affected materially.

If the sheeting is left in place, the coefficient of sliding friction μ' may be reduced, increasing C_d and the load. When the sheeting is retrieved, it should be pulled in

CONCRETE PRESSURE PIPE

TABLE 4.1
Marston Soil Coefficients (C_d) for Trench Conduits

A = K_μ' = •1924 Granular materials without cohesion
B = K_μ' = •165 Maximum for sand and gravel
C = K_μ' = •150 Maximum for saturated top soil
D = K_μ' = •130 Ordinary maximum for clay
E = K_μ' = •110 Maximum for saturated clay

H/B_d	A	B	C	D	E
0.05	0.050	0.050	0.050	0.050	0.050
0.10	0.098	0.098	0.099	0.099	0.099
0.15	0.146	0.146	0.147	0.147	0.148
0.20	0.192	0.194	0.194	0.195	0.196
0.25	0.238	0.240	0.241	0.242	0.243
0.30	0.283	0.286	0.287	0.289	0.290
0.35	0.327	0.331	0.332	0.335	0.337
0.40	0.371	0.375	0.377	0.380	0.383
0.45	0.413	0.418	0.421	0.425	0.428
0.50	0.455	0.461	0.464	0.469	0.473
0.55	0.496	0.503	0.507	0.512	0.518
0.60	0.536	0.544	0.549	0.555	0.562
0.65	0.575	0.585	0.591	0.598	0.606
0.70	0.614	0.625	0.631	0.640	0.649
0.75	0.651	0.664	0.672	0.681	0.691
0.80	0.689	0.703	0.711	0.722	0.734
0.85	0.725	0.741	0.750	0.763	0.775
0.90	0.761	0.779	0.789	0.802	0.817
0.95	0.796	0.816	0.827	0.842	0.857

H/B_d	A	B	C	D	E
3.00	1.780	1.904	1.978	2.083	2.196
3.10	1.810	1.941	2.018	2.128	2.247
3.20	1.840	1.976	2.057	2.172	2.297
3.30	1.869	2.010	2.095	2.215	2.346
3.40	1.896	2.044	2.131	2.257	2.394
3.50	1.923	2.076	2.167	2.298	2.441
3.60	1.948	2.107	2.201	2.338	2.487
3.70	1.973	2.137	2.235	2.376	2.531
3.80	1.997	2.166	2.267	2.414	2.575
3.90	2.019	2.194	2.299	2.451	2.618
4.00	2.041	2.221	2.329	2.487	2.660
4.10	2.062	2.247	2.359	2.522	2.701
4.20	2.082	2.273	2.388	2.556	2.741
4.30	2.102	2.297	2.416	2.589	2.780
4.40	2.121	2.321	2.443	2.621	2.819
4.50	2.139	2.344	2.469	2.652	2.856
4.60	2.156	2.366	2.495	2.683	2.893
4.70	2.173	2.388	2.520	2.713	2.929
4.80	2.189	2.409	2.543	2.742	2.964

TABLE 4.1 (cont)

H/B_d	A	B	C	D	E
1.00	0.830	0.852	0.864	0.881	0.898
1.05	0.864	0.887	0.901	0.919	0.938
1.10	0.897	0.922	0.937	0.957	0.977
1.15	0.929	0.957	0.973	0.994	1.016
1.20	0.961	0.991	1.008	1.031	1.055
1.25	0.992	1.024	1.042	1.067	1.093
1.30	1.023	1.057	1.076	1.103	1.131
1.35	1.053	1.089	1.110	1.139	1.168
1.40	1.082	1.121	1.143	1.173	1.205
1.45	1.111	1.152	1.176	1.208	1.241
1.50	1.140	1.183	1.208	1.242	1.278
1.55	1.167	1.213	1.240	1.276	1.313
1.60	1.195	1.243	1.271	1.309	1.349
1.65	1.221	1.272	1.301	1.342	1.384
1.70	1.248	1.301	1.332	1.374	1.418
1.75	1.273	1.329	1.361	1.406	1.452
1.80	1.299	1.357	1.391	1.437	1.486
1.85	1.323	1.385	1.420	1.469	1.520
1.90	1.348	1.412	1.448	1.499	1.553
1.95	1.372	1.438	1.476	1.530	1.586
2.00	1.395	1.464	1.504	1.560	1.618
2.10	1.440	1.515	1.558	1.618	1.682
2.20	1.484	1.564	1.610	1.675	1.744
2.30	1.526	1.612	1.661	1.731	1.805
2.40	1.567	1.658	1.711	1.785	1.865
2.50	1.606	1.702	1.759	1.838	1.923
2.60	1.643	1.745	1.805	1.890	1.980
2.70	1.679	1.787	1.850	1.940	2.036
2.80	1.714	1.827	1.894	1.989	2.090
2.90	1.747	1.867	1.937	2.037	2.144
4.90	2.204	2.429	2.567	2.770	2.999
5.00	2.219	2.448	2.590	2.798	3.032
5.10	2.234	2.467	2.612	2.825	3.065
5.20	2.247	2.486	2.633	2.851	3.098
5.30	2.261	2.503	2.654	2.877	3.129
5.40	2.273	2.520	2.674	2.901	3.160
5.50	2.286	2.537	2.693	2.926	3.190
5.60	2.298	2.553	2.712	2.949	3.220
5.70	2.309	2.568	2.730	2.972	3.248
5.80	2.320	2.583	2.748	2.995	3.277
5.90	2.330	2.598	2.766	3.017	3.304
6.00	2.340	2.612	2.782	3.038	3.331
6.20	2.360	2.639	2.814	3.079	3.383
6.40	2.377	2.664	2.845	3.118	3.433
6.60	2.394	2.687	2.873	3.155	3.481
6.80	2.409	2.709	2.900	3.190	3.527
7.00	2.423	2.730	2.925	3.223	3.571
7.20	2.436	2.749	2.949	3.255	3.613
7.40	2.448	2.767	2.971	3.285	3.653
7.60	2.459	2.784	2.992	3.313	3.691
7.80	2.470	2.799	3.012	3.340	3.728
8.00	2.479	2.814	3.031	3.366	3.763
8.50	2.500	2.847	3.073	3.424	3.845
9.00	2.517	2.875	3.109	3.476	3.918
9.50	2.532	2.898	3.141	3.521	3.983
10.0	2.543	2.919	3.167	3.560	4.042
15.0	2.591	3.009	3.296	3.768	4.378
20.0	2.598	3.026	3.325	3.825	4.490
30.0	2.599	3.030	3.333	3.845	4.539
40.0	2.599	3.030	3.333	3.846	4.545

TABLE 4.2*
Approximate Maximum Backfill Loads on Trench Conduits — lb/lin-ft

Granular materials without cohesion (Kμ' = 0.1924 and w = 100 pounds per cubic foot)

H†	WIDTH OF TRENCH AT TOP OF CONDUIT (IN FEET)								
	2	3	4	5	6	7	8	9	10
4	500	900	1,300	1,700	2,100	2,500	2,900	3,200	3,600
6	700	1,200	1,800	2,400	2,900	3,500	4,100	4,700	5,300
8	800	1,500	2,200	2,900	3,700	4,400	5,200	6,000	6,800
10	900	1,600	2,500	3,400	4,400	5,300	6,200	7,200	8,200
12	900	1,800	2,800	3,800	4,900	6,000	7,200	8,300	9,700
14	900	1,900	3,000	4,200	5,500	6,700	8,000	9,300	10,600
16	1,000	2,000	3,200	4,500	6,000	7,400	8,800	10,300	11,700
18	1,000	2,100	3,400	4,800	6,300	8,000	9,600	11,200	12,800
20	1,000	2,100	3,500	5,000	6,700	8,400	10,200	12,000	13,800
25	1,000	2,200	3,800	5,500	7,400	9,300	11,600	13,700	16,000
30	1,000	2,300	3,900	5,900	8,000	10,200	12,600	15,100	17,700

*By the Marston formula ($W_d = C_d w B_d^2$); surface loads not included.
†H = depth of fill to top of conduit, in feet.

TABLE 4.2a*

Saturated Top Soil (Kμ' = 0.150 and w = 110 pounds per cubic foot)

H†	WIDTH OF TRENCH AT TOP OF CONDUIT (IN FEET)								
	2	3	4	5	6	7	8	9	10
4	600	1,100	1,500	2,000	2,400	2,800	3,200	3,600	4,000
6	900	1,500	2,100	2,800	3,500	4,100	4,700	5,400	6,000
8	1,000	1,800	2,600	3,500	4,300	5,300	6,200	7,000	7,900
10	1,100	2,100	3,100	4,100	5,200	6,300	7,300	8,400	9,600
12	1,200	2,300	3,500	4,800	6,000	7,300	8,500	9,800	11,100
14	1,300	2,500	3,800	5,300	6,700	8,100	9,600	11,100	12,600
16	1,300	2,600	4,100	5,700	7,300	9,000	10,600	12,300	14,000
18	1,300	2,700	4,400	6,100	7,900	9,700	11,600	13,400	15,400
20	1,400	2,800	4,600	6,500	8,500	10,500	12,500	14,600	16,600
25	1,400	3,000	5,000	7,200	9,500	11,900	14,400	17,100	19,600
30	1,400	3,100	5,300	7,700	10,400	13,200	16,000	19,100	22,000

*By the Marston formula ($W_d = C_d w B_d^2$); surface loads not included.
†H = depth of fill to top of conduit, in feet.

TABLE 4.2b*

Sand and gravel ($K\mu' = 0.165$ and $w = 120$ pounds per cubic foot)

H†	WIDTH OF TRENCH AT TOP OF CONDUIT (IN FEET)								
	2	3	4	5	6	7	8	9	10
4	700	1,100	1,600	2,100	2,600	3,100	3,500	4,000	4,400
6	900	1,500	2,200	2,900	3,600	4,400	5,000	5,800	6,600
8	1,000	1,900	2,800	3,700	4,600	5,500	6,500	7,500	8,400
10	1,200	2,200	3,200	4,300	5,500	6,600	7,800	8,900	10,200
12	1,200	2,400	3,600	4,900	6,300	7,600	9,000	10,400	11,800
14	1,300	2,500	4,000	5,400	7,000	8,500	10,100	11,700	13,300
16	1,300	2,700	4,200	5,900	7,600	9,400	11,100	13,000	14,800
18	1,400	2,800	4,500	6,300	8,200	10,100	12,000	14,100	16,200
20	1,400	2,900	4,700	6,600	8,700	10,800	13,000	15,200	17,400
25	1,400	3,000	5,100	7,300	9,700	12,200	14,900	17,600	20,300
30	1,400	3,100	5,300	7,800	10,600	13,400	16,500	19,500	22,800

*By the Marston formula ($W_d = C_d w B^2{}_d$); surface loads not included.
†H = depth of fill to top of conduit, in feet.

increments of 3-4 ft to allow time for frictional forces to develop between the backfill and the sides of the trench. This should result in the most favorable loading condition for a conduit in a sheeted trench. If the sheeting is pulled as the trench is filled, the value of $K\mu'$ is that for the fill material and the soil of the trench sides, but the value of B_d is that between the back faces of the sheeting. If the sheeting is pulled after all or most of the fill is completed, then the mass of fill material may retain its shape for a time, thus tending to eliminate the frictional load transferences and substantially increase the load on the conduit.

As the trench becomes very wide compared to the outside diameter of the conduit, it approaches the conditions of an embankment conduit.

Embankment Conduits

Embankment conduits are those that are covered by fills or embankments such as railway embankments, highway embankments, and earth dams.

Two classes of embankment conduits. (1) Positive-projecting conduits are embankment conduits that are installed in shallow bedding with the top of the conduit projecting above the surface of the natural ground and are then covered with earth fill, as illustrated in Fig. 4.3. This class also includes those conduits that are installed in trenches wider than two or three times the maximum outside diameter of the conduit. The trench width that gives the same loading as the

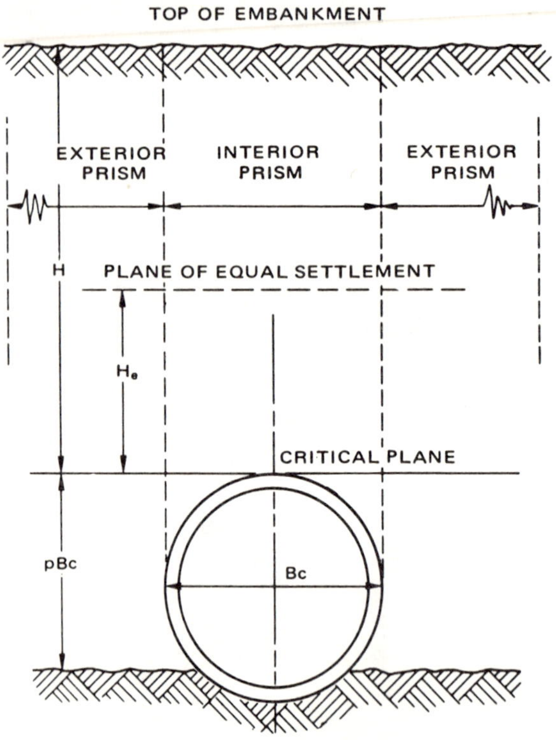

Fig. 4.3 Positive-projecting conduit

projecting conduit load is called the transition trench width. [See section on "Trench Conduits."] (2) Negative-projecting conduits are embankment conduits that are installed in relatively narrow trenches where the tops of the conduits are below the level of the natural ground surface, as illustrated in Fig. 4.4. The final grade above the conduits is appreciably higher than natural ground.

Loads on positive projecting conduits. When considering loads on embankment conduits, the prism of backfill directly above the pipe and bounded by vertical planes tangent to the sides of the pipe is the interior prism. The exterior prisms are the prisms of backfill adjacent to the vertical planes on both sides of the pipe and of indefinite width. The load transmitted to the top of the pipe will be equal to the weight of the interior prism of soil plus or minus the total friction force that develops along the two vertical planes bounding the interior prism. Unless the embankment material on each side of the pipe is thoroughly compacted, there is a tendency for the exterior prisms of soil, because they are longer, to compress more than the interior prism. The friction forces on the vertical tangent planes will then act to make the load on the pipe greater than the weight of the interior prism of soil.

However, if the pipe is placed on a slightly yielding foundation and the compacted embankment material on each side of the pipe is placed on a firm foundation, then, as fill is built up, the pipe will settle more than the adjacent soil. This will reverse the friction forces and make the load on the pipe less than the weight of the soil in the interior prism.

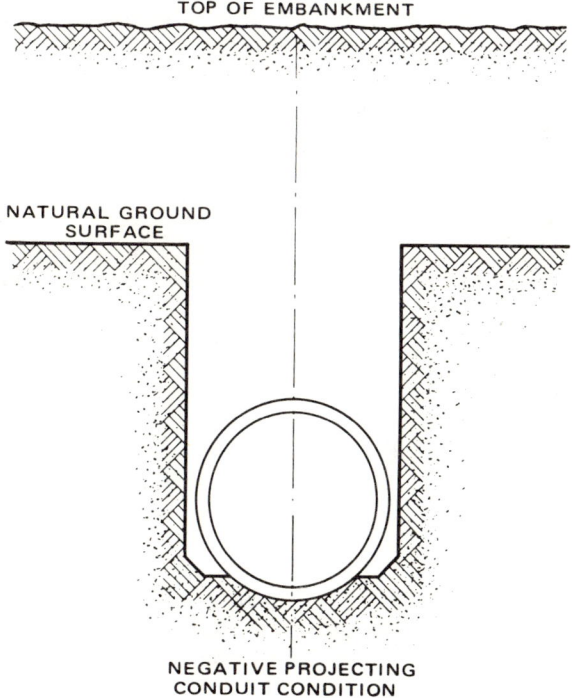

Fig. 4.4 **Negative-projecting conduit**

Load formula — positive projecting conduits. The load on a positive-projecting conduit is computed by the Marston equation

$$W_c = C_c w B_c^2$$

where W_c = load per unit of length; C_c = load calculation coefficient; w = unit weight of fill material; and B_c = outside diameter of the pipe.

Outside pipe diameter. Of the three factors in the previous equation on which the load depends, the horizontal breadth of the conduit B_c is the most easily determined, and it has a constant value throughout the life of the structure.

Significance of soil density. Soil materials differ widely in density characteristics, and their unit weights may vary from about 70 lb/cu ft up to a maximum of 150 lb/cu ft. The upper limit of this range has been increased significantly by the application of improved methods of soil compaction in earthwork construction operations. It is evident that, because the load on an underground conduit is directly dependent upon the unit weight of the embankment material, the height of fill which will produce a given load will vary considerably when composed of different kinds of soil or soils in different states of compaction. Unit weights of most soil embankments fall within the range of 100–135 lb/cu ft.

Load coefficient. The load on a rigid pipe also depends on the load coefficient C_c. This coefficient is dependent upon several physical factors: (1) the ratio of the height of fill to the horizontal breadth of the conduit, H/B_c, (2) the coefficient of internal friction of the soil, μ; (3) the projection ratio, p; (4) the settlement ratio, r_{sd}.

The ratio H/B_c. It is obvious that, because the value of B_c is fixed for any particular rigid pipe conduit, the ratio H/B_c will vary with the height of fill H. The height of fill can be definitely measured in the case of a completed structure or determined from the plans of a proposed structure. There are, therefore, no particular difficulties encountered in the determination of the ratio H/B_c. In most cases, the designer will have a definite height of fill given and will need to determine the load produced by the fill and the strength of pipe required to carry the load. In some cases, he may have a definite strength of pipe specified and will want to find a height of fill that will not produce a load on the pipe greater than that which it is capable of supporting.

Coefficient of internal friction. The second factor under consideration is the coefficient of internal friction μ of the embankment soil. Like the unit weight of soil, the internal friction coefficient may vary over a considerable range of values, perhaps from 0.3 to 1.0. Fortunately, this property of soil does not influence the load on a pipe in direct proportion, and the relationship is such that relatively wide variations in μ cause only minor variations in load on the structure. In the case of projecting conduits it is reasonable to use a single value of μ for design purposes, regardless of the type of soil. The value $\mu = 0.6$ is suggested as a good average figure which will give results that are on the side of safety, but which are neither ultra-safe nor uneconomical.

Projection ratio. The third physical factor upon which the load coefficient depends is the projection ratio p, the vertical distance between the top of a pipe and the natural ground surface adjacent thereto divided by the breadth of the pipe B_c. It is easily and definitely determinable in cases where the natural ground surface is fairly level for a considerable distance on each side of the conduit. Where the natural ground surface slopes toward or away from the conduit, the average vertical distance over a horizontal distance approximately equal to B_c on each side of the conduit can be used as a basis for computing the projection ratio.

Settlement ratio. The fourth factor which determines the load coefficient is the settlement ratio r_{sd}. This factor determines the direction of action of the friction forces. The settlement ratio has been measured for relatively few projecting type installations. However, on the basis of these measurements, the following tentative values of settlement ratio are recommended for use in the design of concrete pipe positive-projecting conduits:

> On rock or unyielding soil $\quad r_{sd} = +1.0$
> On ordinary soil $\quad r_{sd} = +0.5$ to $+0.8$
> On yielding foundations $\quad r_{sd} = 0.0$ to $+0.5$

The settlement ratio can be negative where the conduit settles more than the adjacent ground. A positive value of settlement ratio indicates a load on the pipe which is greater than the weight of the prism of soil directly above it. A negative value indicates a load on the pipe which is less than the weight of the prism of soil above it.

Determination of load coefficient C_c. An examination of the load formulas and of the load computation diagram, Fig. 4.5, reveals that the projection ratio and the settlement ratio influence the load on a conduit in the relationship of the product of the two factors r_{sd} and p. Where concrete pipe is installed as a projecting conduit, both factors usually have a positive value less than unity, and the product of the two is therefore less than either factor. Figure 4.5 enables the designer to determine values of C_c for various values of settlement ratio r_{sd}, projection ratio p, and H/B_c.

Negative-projecting conduits. Negative-projecting conduits are those installed in shallow trenches where the top of the conduit is below the natural ground level and an embankment is built over it, as illustrated in Fig. 4.4. Sometimes a conduit is located in higher ground to one side of the natural watercourse, and the stream flow is diverted to the new location. The conduit is often laid in a relatively narrow trench dug in the undisturbed soil of the side hill. If the depth of the trench is such that the top of the pipe is below natural ground, the conduit is considered to be a negative-projecting embankment conduit. In the case of negative-projecting conduits, the load transmitted to the pipe equals the weight of the interior prism of soil above the pipe plus or minus friction forces along the side of that prism, as in the case of positive-projecting conduits.

Load formula — negative-projecting conduits. The procedure for computing the load on a negative-projecting conduit is similar to the procedure for positive-projecting conduits. The load is computed by the equation

$$W_c = C_n w\, B_d^2$$

where C_n = the load coefficient for negative-projecting conduits.

The values of C_n are dependent on the product of the projection ratio p, and the settlement ratio r_{sd}, as well as on the ratio of the height of fill to the width of the trench, H/B_d. C_n is taken from the appropriate chart when other values have been determined.

Projection ratio for negative-projecting conduits. The projection ratio p' is the distance from the top of the conduit to the natural ground surface divided by the width of the trench. It is comparatively easy to determine, even when the natural ground surface is on a transverse slope. The vertical distance involved may be taken

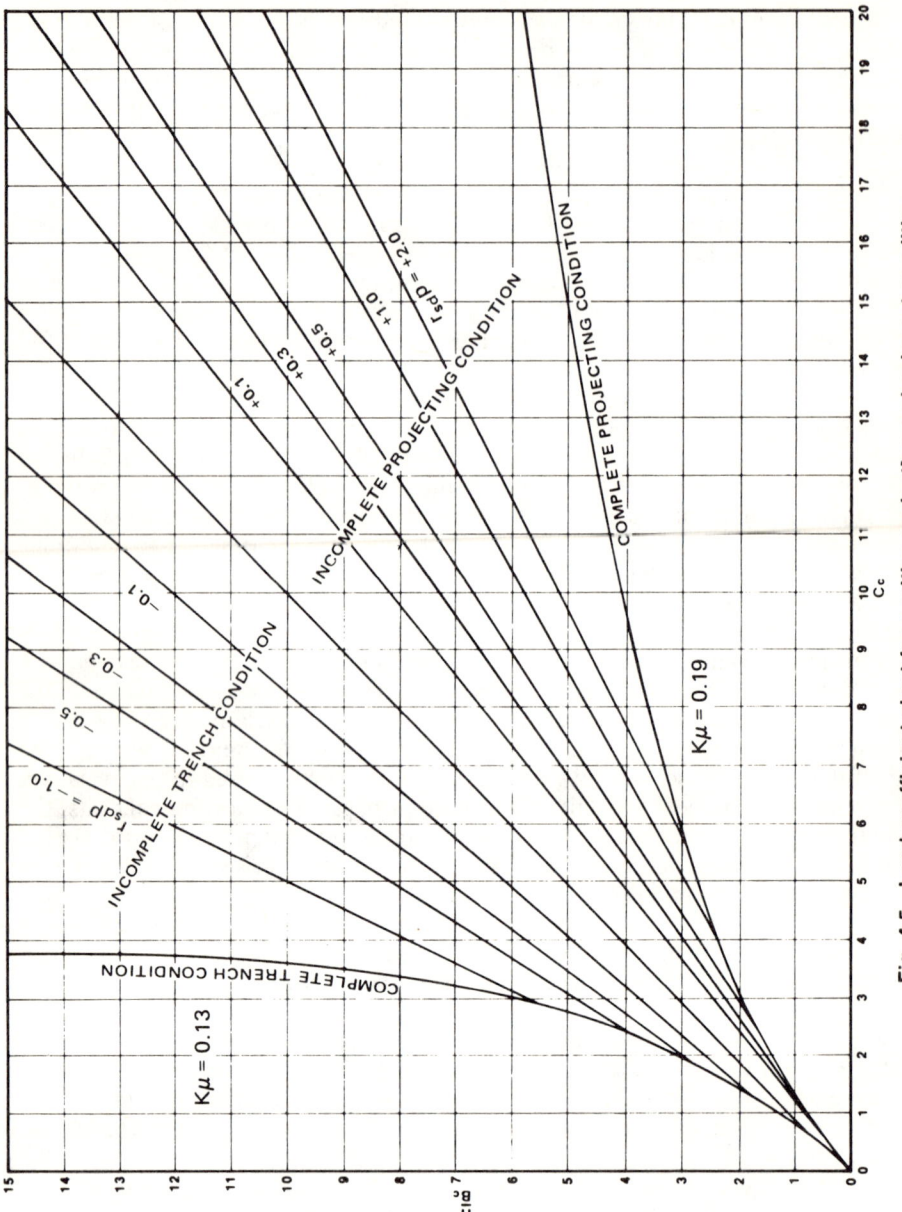

Fig. 4.5 Load coefficient chart for positive projecting embankment condition

as the average of the distances from the top of the conduit to the ground surface at both sides of the trench.

Negative settlement ratio. Since the backfill over the pipe is normally more compressible than the natural soil in which the trench is dug, the interior prism tends to settle more than the exterior prism on each side. Therefore, the numerical value of the settlement ratio r_{sd} will be negative. A negative settlement ratio indicates that the load on the pipe will be less than the actual weight of the interior prism of soil. This reduction in load may be great enough to permit the construction of much higher fills over the pipe.

To understand this design factor more easily, it is necessary to have an understanding of the manner in which the soil over the pipe acts in settling. In developing his equations, M.G. Spangler of Iowa State University considered that a negative-projecting installation was either a "complete trench condition" where the relative settlement between the interior and exterior prisms are such that the theoretical plane of equal settlement is above the top of the embankment, or an "incomplete trench condition" where that plane of equal settlement is at some elevation below the top of the embankment. Examples are illustrated in Fig. 4.6.

In the first case, upward shearing stresses on the interior prism act throughout the complete height of the fill. In the second case, they act only throughout the distance from the top of the conduit to the plane of equal settlement. From there to the top of the embankment, the additional load on the pipe is equal to the weight of the interior prism of soil in that region.

The values of the settlement ratio in this study vary between zero and -2.0 for the incomplete trench. W.J. Schlick, also of Iowa State University, has stated that he believes values of -1.0 to -2.0 can be obtained for $r_{sd}p'$. It is probable that the value of r_{sd} in normal installations will fall between zero and -1.0, with -0.5 as a reasonable figure to use. This means that the trench is deep enough for some upward shearing forces to develop and that the soil under the conduit will yield slightly over a length of time.

Figures 4.7-4.10 were prepared by M.G. Spangler and were first presented to the Highway Research Board in 1950. Extensive mathematical computations were made in their preparation. The formulas derived are not presented here.

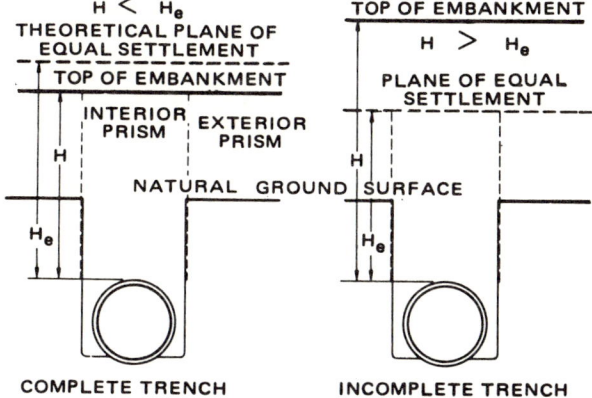

Fig. 4.6 Complete and incomplete trench conditions

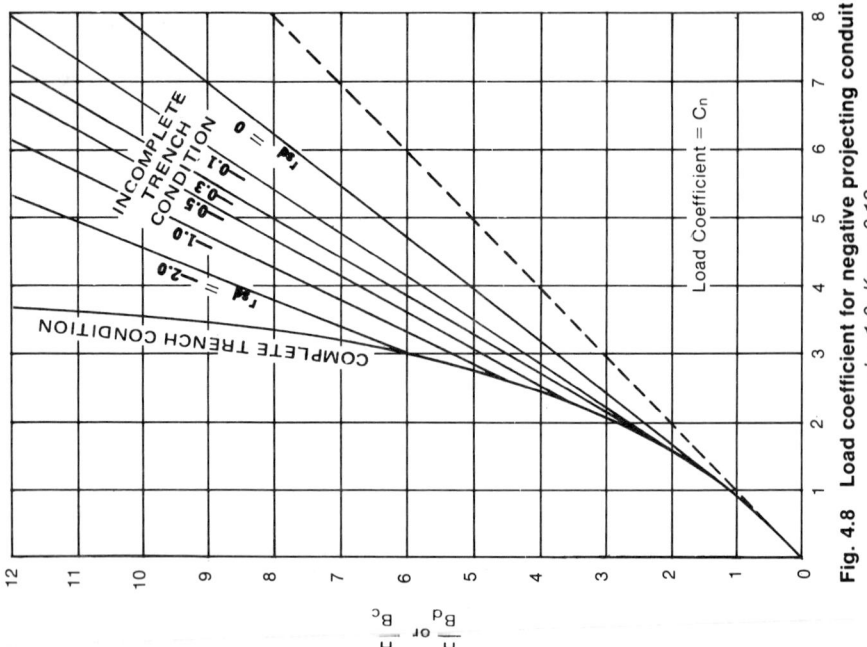

Fig. 4.8 Load coefficient for negative projecting conduit
$p' = 1.0$, $K\mu = 0.13$

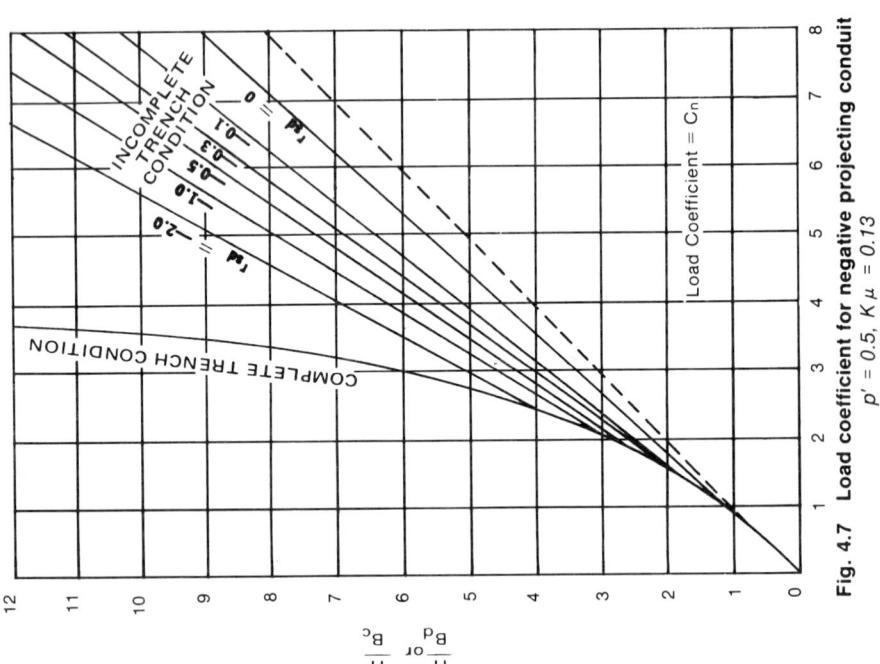

Fig. 4.7 Load coefficient for negative projecting conduit
$p' = 0.5$, $K\mu = 0.13$

EXTERNAL LOADS

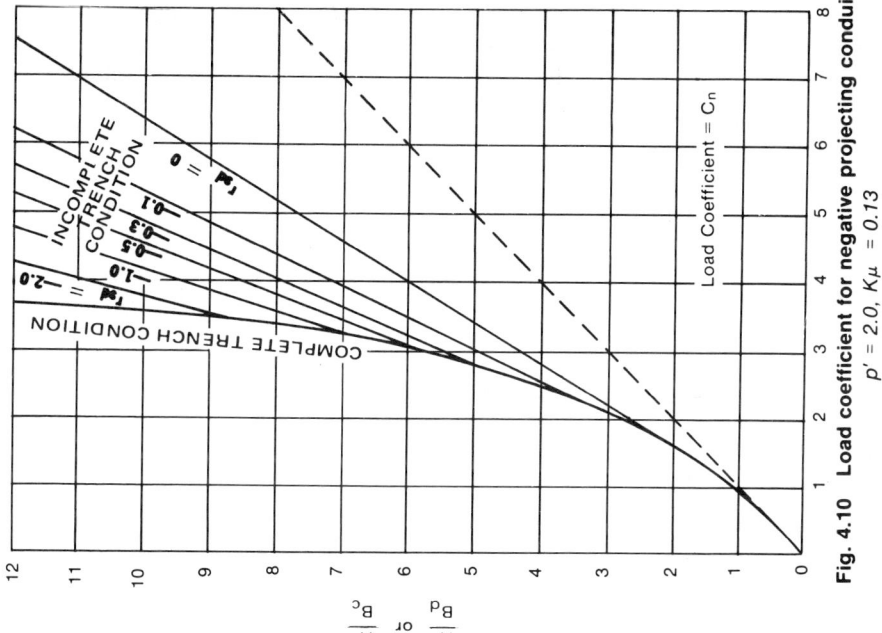

Fig. 4.10 Load coefficient for negative projecting conduit
$p' = 2.0, K\mu = 0.13$

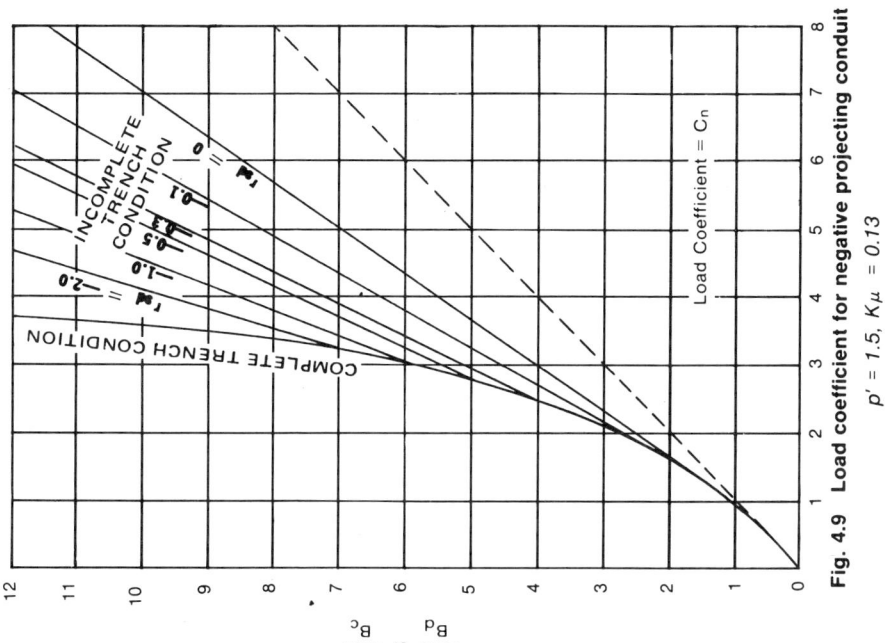

Fig. 4.9 Load coefficient for negative projecting conduit
$p' = 1.5, K\mu = 0.13$

The diagrams showing values of C_n versus H/B_d for various values of r_{sd} have been drawn for projection-ratio values of $p' = 0.5, 1.0, 1.5,$ and 2.0. For other values of p' between zero and 2.0, interpolation will give values of C_n with sufficient accuracy for design purposes. The coefficient of internal friction μ was assumed to be 0.2 ($K\mu = 0.13$) in the construction of the diagrams. This is believed to be a safe design value. These diagrams cannot be used for determination of loads involving coefficients of internal friction other than $\mu = 0.2$.

The induced trench. In the early days of Marston's research on conduit loads, he was impressed by the very high loads that can develop on positive-projecting conduits when the conditions are such that large shearing forces are added to the weight of the interior prism of soil directly over the conduit. He worked to devise a method of construction that would reduce or eliminate these shearing forces, or possibly even reverse their direction to provide a benevolent action as in the case of trench conduits. With this objective, he developed the method of construction which creates the "induced trench." The soil on both sides and above the conduit is thoroughly compacted by rolling and tamping, and the trench is dug in this compacted fill by removing the prism of material directly over the conduit. This trench is refilled with very loose compressible material, after which the embankment is completed in a normal manner, as illustrated in Fig. 4.11. The object of this method of construction is to create a condition wherein the interior prism of material will settle more than the adjacent exterior prisms. Therefore, the trench in the artificially compacted material must be deep enough and the backfill material loose enough to ensure this action.

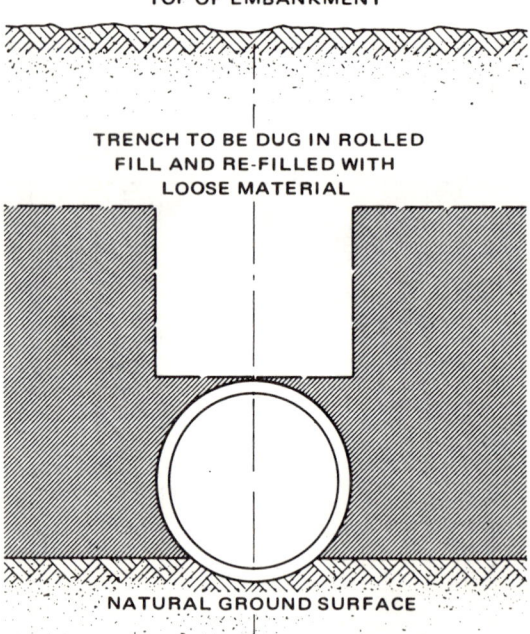

Fig. 4.11 Induced trench conduit

Application of the induced trench. The preceding theoretical analysis of loads on negative-projecting conduits can be used to estimate loads on conduits installed by the induced trench method. When used for this purpose, the top surface of the initially compacted material can be considered to be the natural ground surface. The width of the trench in the induced trench method can be made much smaller than the negative-projecting case, and, for the most favorable results, it should be made no wider than the outside diameter of the conduit B_c.

Several states now recognize the advantages to be gained by the negative-projecting and the induced trench methods of installation. California, Washington, Alabama, Texas, North Dakota, and North Carolina have used these methods in constructing pipelines under high fills.

To begin the California method, the new embankment is constructed to an elevation above the top of the conduit which is equal to the external diameter of the conduit. The trench is then excavated and the pipe is installed. After installation, backfill material is placed. Backfill to the top of the conduit should be granular material approved by the engineer and should be compacted. Backfill material from the top of the pipe to a point above the pipe equal to its external diameter should be placed in the loosest possible condition. At the time backfill material is placed above the pipe, the vertical surfaces of the trench above the top of the pipe should not extend more than 8 in. beyond the outside dimensions of the pipe.

These methods provide an installation with a generous factor of safety.

Determination of Live Load

In the selection of pipe, it is necessary to evaluate the effect of live loads. Live load considerations are necessary in the design of pipe installed with shallow cover under railroads, airports, and unsurfaced highways. The distribution of a live load at the surface on any horizontal plane in the subsoil is shown in Fig. 4.12. The intensity of the load on any plane in the soil mass is greatest at the vertical axis directly beneath the point of application and decreases in all directions outward from the center of application. As the distance between the plane and the surface increases, the intensity of the load at any point on the plane decreases.

Highway live loads. If a rigid or flexible pavement designed for heavy-duty traffic is provided, the intensity of a truck wheel load is usually reduced sufficiently so that the live load transmitted to the pipe is negligible. In the case of flexible pavements designed for light-duty traffic but subjected to heavy truck traffic, the flexible pavement should be considered as fill material over the top of the pipe.

The total live load transmitted to a pipe underground can be determined by calculating the volume of the pressure intensity diagram shown in Fig. 4.13. The volume is closely approximated by an elliptical cylinder and ellipsoid. Based on this type of loading configuration, the total live load per linear foot of pipe is computed by the equation

$$W_L = \frac{\pi W L \, (2p_1 + p_2)}{L + 24}$$

The pressure intensities p_1 and p_2 are determined by the Boussinesque solution for stresses in a semi-infinite elastic medium.

The area of any horizontal plane in the subsoil subjected to a truck load can be defined by planes descending from the edge of the contact area at the surface outward at a 30 deg angle with the vertical. Based on a typical highway truck loading with dual tires, the limitations of the stress distribution for various heights of cover are shown in Fig. 4.14.

The length and width of the loaded area is given by the following equation, with the width limited to a maximum dimension equal to the outside diameter of pipe B_c.

$$L = 1.0 + 1.155 \, H$$

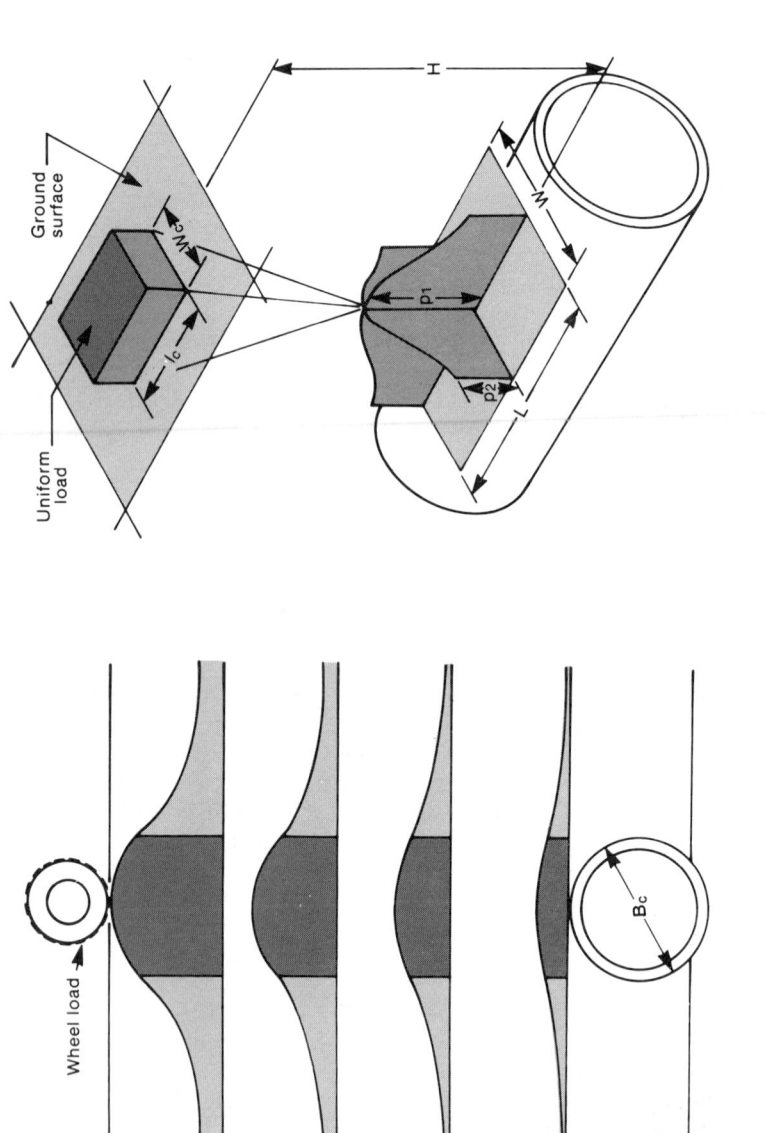

Fig. 4.13 Highway live load pressure distribution

Fig. 4.12 Live load distribution

Table 4.3 presents total live loads for circular pipe, expressed in pounds per linear foot. The tables are based on a 16,000-lb, dual-tired wheel load (32,000-lb axle load) and 80-psi tire pressure. Recommended impact factors to be used in designing pipe with less than 3 ft of cover when subjected to dynamic traffic loads are listed in Table 4.4.

Construction loads. During grading operations it may be necessary for heavy construction equipment to travel over installed pipe. Unless adequate protection is provided, the pipe may be subjected to load concentrations in excess of the design loads. Before heavy construction equipment is permitted to cross a pipe, a temporary earth fill should be constructed to an elevation of at least 3 ft over the top of the pipe. The fill should be of sufficient width to prevent possible lateral displacement of the pipe.

TABLE 4.3

Highway Live Loads for Circular Pipe—lb/lin-ft

PIPE SIZE D IN INCHES	\multicolumn{10}{c}{HEIGHT OF FILL H ABOVE TOP OF PIPE IN FEET}	PIPE SIZE D IN INCHES
	0.5	1.0	1.5	2.0	2.5	3.0	3.5	4.0	5.0	6.0	
12	4849	2730	1524	954	574	361	316	237	168	46	12
15	5915	3322	1872	1163	700	441	358	289	205	56	15
18		3926	2196	1369	825	520	454	341	242	66	18
21		**4498**	2525	1580	950	599	522	392	279	77	21
24			2856	1788	1077	679	592	444	316	87	24
27			**3096**	1998	1201	757	660	496	353	97	27
30				2204	1328	837	731	548	390	107	30
33				**2370**	1452	915	798	599	426	117	33
36					1580	996	869	652	463	127	36
42					**1679**	1154	1007	755	537	147	42
48						**1221**	1146	859	611	168	48
54							**1185**	962	684	188	54
60								**998**	758	208	60
66									832	228	66
72									**857**	249	72
78										269	78
84										275	84

Unsurfaced Roadway; 16,000-Pound Wheel Load; Dual Tires; 80 p.s.i. Tire Pressure; Impact Included. Last number in each column (**bold type**) indicates maximum load for any given fill height. Interpolate for intermediate fill heights.

TABLE 4.4

*Impact Factors for Highway Truck Loads**

Height of Cover — H	Impact Factor — I_f
0 ft-0 in. to 1 ft-0 in.	1.3
1 ft-1 in. to 2 ft-0 in.	1.2
2 ft-1 in. to 2 ft-11 in.	1.1
3 ft-0 in. and greater	1.0

*Impact factors recommended by the American Association of State Highway Transportation Officials in *Standard Specifications for Highway Bridges*.

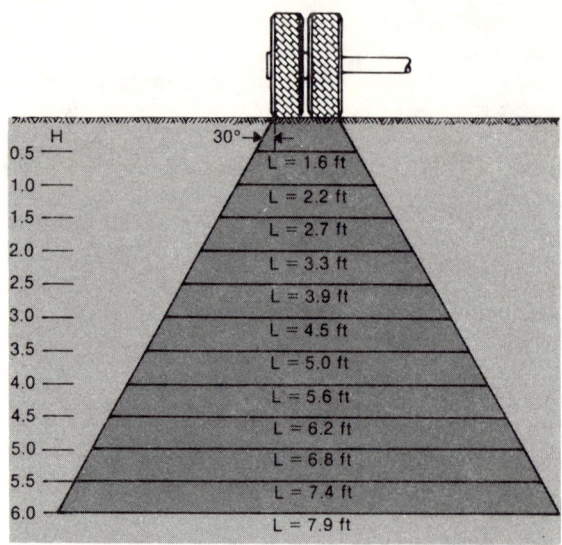

Fig. 4.14 Limiting load distribution for typical dual-tire truck loading

Bibliography

1. *Concrete Culverts and Conduits*. Portland Cement Association.
2. PECKWORTH, H.F. *Concrete Pipe Handbook*. American Concrete Pipe Association (1965).
3. *Concrete Pipe Design Manual*. American Concrete Pipe Association (1974).
4. *Concrete Sewers*. Portland Cement Association (1951).
5. MARSTON, ANSTON, The Theory of External Loads on Closed Conduits in the Light of Latest Experiments. Bulletin 96, Iowa State Engineering Experiment Station, Iowa State University (1930).
6. Handbook of Culvert and Drainage Practice. ARMCO Drainage & Metal Products, Inc. (1948).
7. *Installation of Concrete Pressure Pipe, Rigid Design*. Ameron.
8. *Installation of Concrete Pressure Pipe, Semi-Rigid Design*. Ameron.
9. Structural Design of Underground Conduits. Technical Release No. 5, US Dept. of Agriculture, Soil Conservation Service (1958).
10. BABCOCK, D.P. Simplified Design Methods for Reinforced Concrete Pipe Under Earth Fills. Proceedings of Highway Research Board (1955).
11. Soils and Pipe Backfill. Engineering Topics No. 5, American Pipe and Construction Co. (1964).
12. SPANGLER, M.G. & HANDY, R.L. *Soil Engineering*. Intext Educational Publishers, 3rd ed. (1973).
13. Reinforced Concrete Pipe Culverts. US Dept. of Commerce, Bureau of Public Roads (1963).
14. SPANGLER, M.G. Underground Conduits, An Appraisal of Modern Research. *Trans. ASCE*, Paper No. 2337, Vol. 113 (1968).
15. Vertical Pressure on Culverts Under Wheel Loads On Concrete Pavement Slabs. No. ST65, Structural Bureau, Portland Cement Association (1944).

CHAPTER 5

Bedding and Backfilling

General

The bedding of pipe has an important effect on its external load-carrying capacity and can be a controlling factor in the design of the pipe. When designing concrete pressure pipe, emphasis is placed on the bedding angle, which determines how the external load will be distributed on the pipe, and thereby influences pipe strength requirements.

Rigid Pipe

In general, it is more economical to design rigid pipes (C300, C301, C302) to accommodate external loading with minimal bedding support than it is to require that the pipe be installed with highly compacted backfill. It is important to avoid laying the pipe on a hard or unyielding surface (such as rock, hard clay, or shale). Usually, fines from the excavation can be placed in the trench bottom to provide sufficient support for the pipe. Where appropriate material is not available from the excavation it may be necessary to import material to provide a cushion.

The Type 3 bedding condition in Fig. 5.1 and 5.2 illustrates the commonly specified requirements for bedding and backfilling of rigid pipe. Although Type 3 bedding is the most frequently specified, it represents a more rigorous bedding design than is actually required to support concrete pipe at moderate covers. For many years less stringent requirements have been used in the field with great success. When rigid concrete pipelines are subjected to moderate earth covers (6 ft or less) the Type 2 bedding details illustrated in Fig. 5.1 are a more accurate representation of the support that is needed without significantly increasing pipe design or installation costs. For severe external loading situations, the Type 4 and Type 5 details illustrated in Fig. 5.1 and 5.2 represent beddings that provide additional pipe support. Engineers should consider these conditions and take advantage of the most economical method.

Load factors. For each type of bedding, a load factor is recommended for rigid pipe. The load factor is the ratio of the strength of the pipe under the installed condition of loading and bedding to the strength of the pipe in the three-edge bearing test.

CONCRETE PRESSURE PIPE

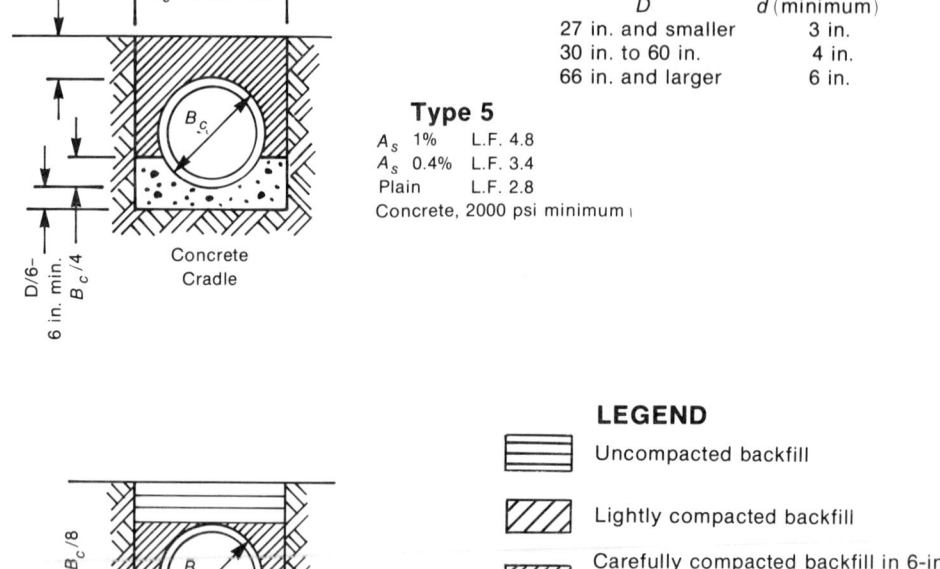

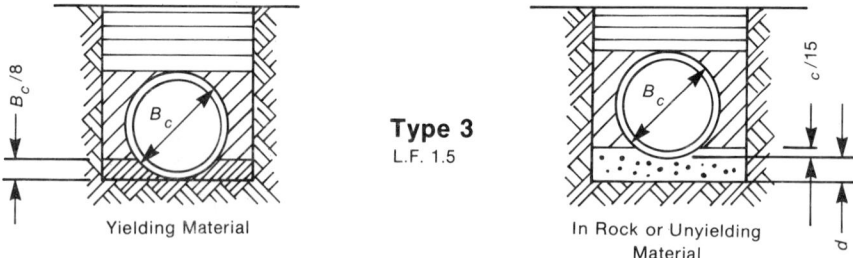

Fig. 5.1 Trench beddings for circular pipe

NOTE: *Inverted cradle may be used to improve load carrying capacity of existing installations.*

BEDDING AND BACKFILLING 53

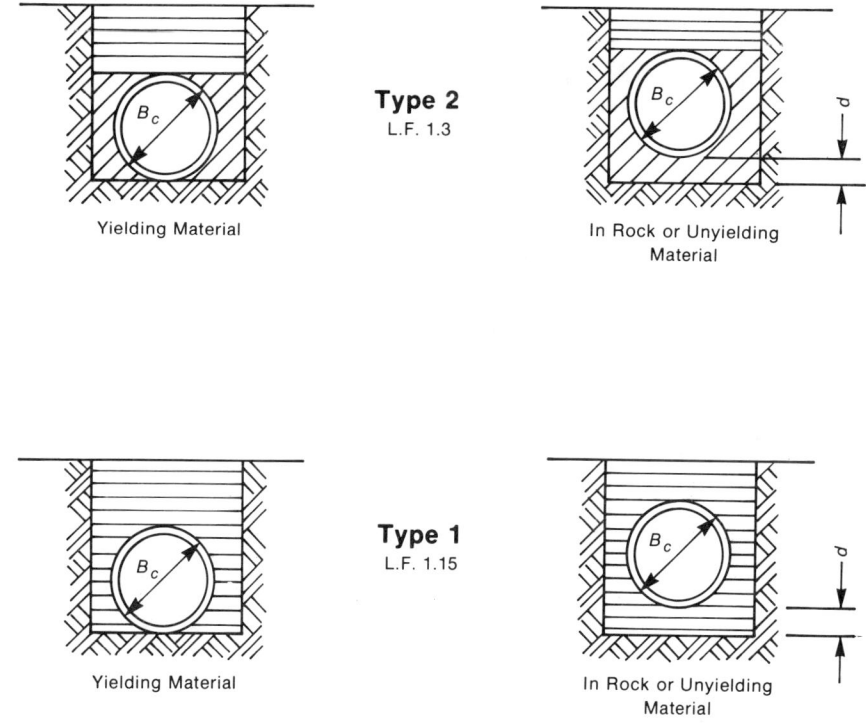

Fig. 5.1 (cont)

Values for load factors are listed in Fig. 5.1 for trench beddings and in Table 5.1 for embankment beddings. For further information on computation of embankment load factors, refer to "Soil Engineering," by M.G. Spangler.[1]

Semi-Rigid Pipe

Concrete pressure pipe manufactured in accordance with AWWA C303 is considered semi-rigid in larger diameters because it develops its ability to support external loads both from its inherent strength and from the support of the surrounding soil as it deflects.

The smaller diameters of C303 pipe may be classified as rigid since they have sufficient strength to support large external loads without injurious deflection or dependence on support from surrounding soil. For such pipe the bedding methods shown in Fig. 5.1 may be utilized. As the diameter becomes larger, support from the surrounding soil becomes increasingly important, as demonstrated in Table. 5.2.

Bedding angle. The bedding angle, which is the angle of uniform support beneath the pipe, is an important design factor that reduces the unit vertical reaction load on the pipe as the angle is increased.*

---

*For a discussion of the correlation between bedding angle and bedding constant k, see Bull. 153 (December 1941), Iowa State Experiment Station, by M.G. Spangler.

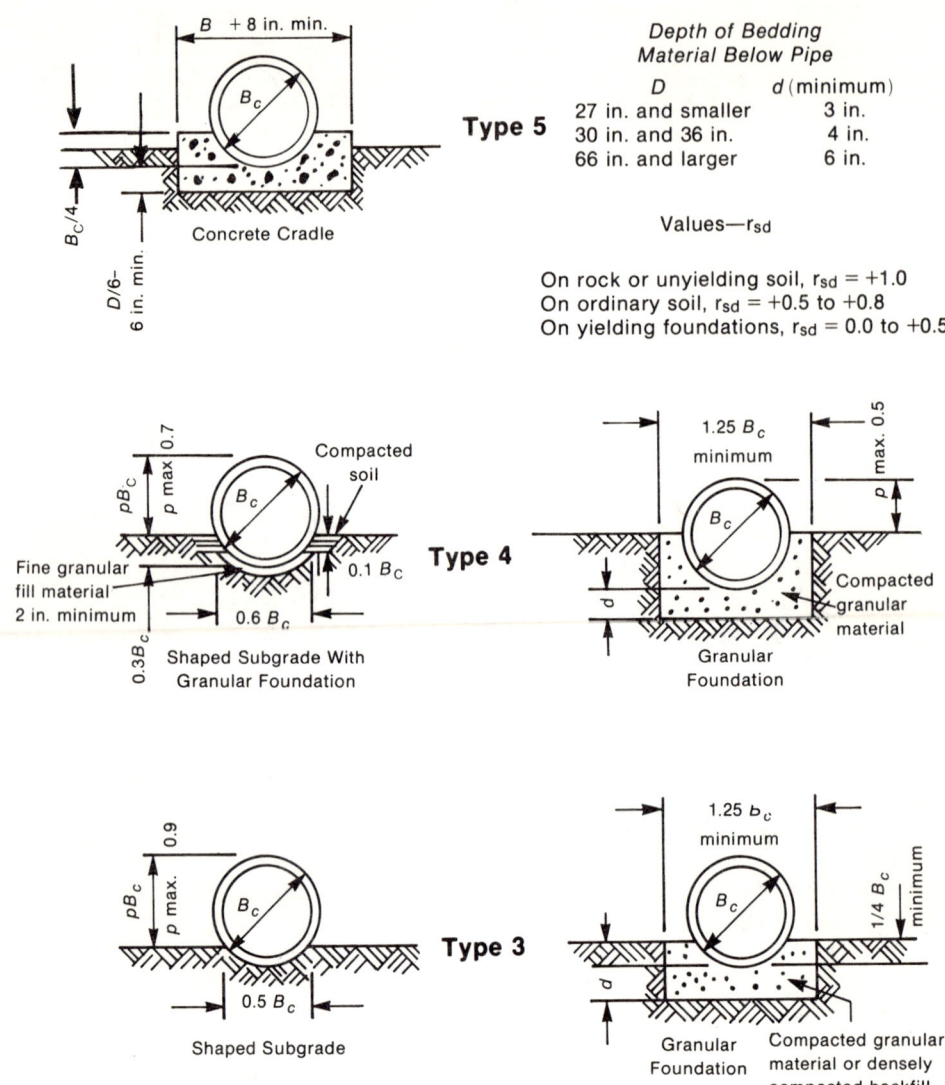

Fig. 5.2 Embankment beddings for circular pipe

NOTE: **Rock or other noncompressible foundation:** *Where ledge rock, rocky or gravelly soil, hard pan or other unyielding foundation material is encountered, the hard unyielding material should be excavated below the elevation of the concrete cradle (Type 5) or the bottom of the pipe bell (Type 4 and Type 3 beddings) for a depth of at least 6 in. or 1/2 in. for each foot of fill over the top of the pipe, whichever is greater, but not more than one quarter of the nominal diameter of the pipe. The width of the excavation should be 1 ft greater than the outside diameter of the pipe. The excavation should be refilled with selected fine compressible material such as silty clay or loam, lightly compacted and shaped as required for the specified class of bedding.*

BEDDING AND BACKFILLING

Lateral support. The lateral support provided by the soil is another major consideration in determining the performance of semi-rigid pipe in resisting external loads. The amount of lateral support which a particular soil is capable of providing is dependent upon the soil's intrinsic physical properties, and upon the degree to which the soil can be compacted.

TABLE 5.1
Load Factors for Circular Pipe
Positive Projecting Embankment Installations

$\frac{H}{B_c}$	Type 3 Bedding					Type 4 Bedding				Type 5 Bedding				
							$p = 0.9$							
	$r_{sd}p=0$	0.1	0.3	0.5	1.0	$r_{sd}p=0$	0.1	0.3	0.5	$r_{sd}p=0$	0.1	0.3	0.5	1.0
0.5	3.01	2.82	2.82	2.82	2.82					11.26	8.87	8.87	8.87	8.87
1.0	2.55	2.35	2.35	2.35	2.35					6.61	5.37	5.37	5.37	5.37
1.5	2.42	2.26	2.16	2.16	2.16	Maximum				5.81	4.83	4.47	4.47	4.47
2.0	2.37	2.20	2.14	2.10	2.10	Recommended				5.48	4.49	4.35	4.19	4.19
3.0	2.31	2.17	2.10	2.07	2.02	Projection Ratio of 0.7				5.18	4.50	4.21	4.06	3.88
5.0	2.27	2.14	2.08	2.04	2.00					4.97	4.37	4.11	3.97	3.81
10.0	2.24	2.12	2.06	2.03	1.99					4.82	4.28	4.04	3.90	3.76
15.0	2.23	2.10	2.05	2.02	1.98					4.77	4.25	4.01	3.88	3.74
							$p = 0.7$							
	$r_{sd}p=0$	0.1	0.3	0.5	1.0	$r_{sd}p=0$	0.1	0.3	0.5	$r_{sd}p=0$	0.1	0.3	0.5	1.0
0.5	2.35	2.27	2.27	2.27	2.27	3.00	2.88	2.88	2.87	7.52	6.54	6.54	6.54	6.54
1.0	2.18	2.08	2.08	2.08	2.08	2.73	2.58	2.58	2.58	5.61	4.79	4.79	4.79	4.79
1.5	2.13	2.03	1.99	1.99	1.99	2.65	2.50	2.44	2.44	5.17	4.46	4.19	4.19	4.19
2.0	2.10	2.01	1.97	1.95	1.95	2.61	2.48	2.42	2.39	4.98	4.35	4.11	3.99	3.98
3.0	2.08	2.00	1.96	1.94	1.91	2.58	2.45	2.40	2.36	4.80	4.25	4.02	3.90	3.75
5.0	2.06	1.98	1.95	1.93	1.90	2.55	2.43	2.38	2.35	4.66	4.18	3.95	3.84	3.70
10.0	2.05	1.98	1.94	1.92	1.89	2.53	2.42	2.36	2.33	4.57	4.12	3.91	3.79	3.66
15.0	2.04	1.97	1.94	1.91	1.89	2.52	2.41	2.36	2.33	4.53	4.09	3.89	3.77	3.65
							$p = 0.5$							
	$r_{sd}p=0$	0.1	0.3	0.5		$r_{sd}p=0$	0.1	0.3	0.5	$r_{sd}p=0$	0.1	0.3	0.5	
0.5	1.94	1.92	1.92	1.92		2.37	2.33	2.33	2.33	4.84	4.54	4.55	4.55	
1.0	1.90	1.86	1.86	1.86		2.31	2.25	2.25	2.25	4.33	3.97	3.97	3.97	
1.5	1.88	1.85	1.83	1.83		2.28	2.23	2.20	2.20	4.18	3.83	3.68	3.68	
2.0	1.88	1.84	1.83	1.82		2.27	2.22	2.20	2.19	4.11	3.79	3.65	3.58	
3.0	1.87	1.84	1.82	1.81		2.26	2.22	2.19	2.18	4.04	3.75	3.62	3.54	
5.0	1.86	1.83	1.82	1.81		2.26	2.21	2.19	2.17	3.99	3.72	3.58	3.51	
10.0	1.86	1.83	1.81	1.80		2.25	2.20	2.18	2.17	3.95	3.69	3.56	3.49	
15.0	1.86	1.83	1.81	1.80		2.25	2.20	2.18	2.17	3.94	3.68	3.56	3.48	
							$p = 0.3$							
	$r_{sd}p=0$	0.1	0.3			$r_{sd}p=0$	0.1	0.3		$r_{sd}p=0$	0.1	0.3		
0.5	1.76	1.76	1.76			2.11	2.10	2.10		3.49	3.41	3.41		
1.0	1.76	1.75	1.75			2.10	2.08	2.08		3.40	3.28	3.28		
1.5	1.75	1.74	1.74			2.09	2.08	2.07		3.37	3.25	3.20		
2.0	1.75	1.74	1.74			2.09	2.08	2.07		3.35	3.24	3.20		
3.0	1.75	1.74	1.74			2.09	2.08	2.07		3.34	3.23	3.18		
5.0	1.75	1.74	1.74			2.09	2.08	2.07		3.33	3.22	3.17		
10.0	1.75	1.74	1.74			2.09	2.08	2.07		3.32	3.22	3.17		
15.0	1.75	1.74	1.74			2.09	2.08	2.07		3.32	3.22	3.17		
							Zero Projecting							
		1.70					2.02				2.83			

Soil compaction. Much remains to be learned about the influence of soil compaction on semi-rigid pipe deflection. When specifying the amount of compaction required, it is of vital importance to consider what degree of soil compaction is economically obtainable in the field for a particular soil. Specifying an unobtainable soil compaction value can result in inadequate support and injurious deflection. Therefore, a conservative assumption of the supporting capability of a soil is recommended. Good field inspection should be provided and

TABLE 5.2
Allowable External Load Bedding Angle 90 deg
For AWWA C303 Type Pipe*

Pipe I.D. In.	Pipe Class	Cylinder Gage	Rod Size	Wrap As/ft in.	$E' = 400$† w kips/ft	$E' = 400$† h ft	$E' = 700$† w kips/ft	$E' = 700$† h ft
10	230	16	7/32	0.23	6.527	100.0+	6.584	100.0+
12	196	16	7/32	0.23	5.893	100.0+	5.976	100.0+
14	171	16	7/32	0.23	4.942	59.2	5.055	100.0+
16	150	16	7/32	0.23	4.961	31.7	5.108	35.8
18	145	15	7/32	0.23	6.086	65.2	6.272	100.0+
20	131	15	7/32	0.23	5.713	29.2	5.943	33.3
21	125	15	7/32	0.23	5.551	24.7	5.805	27.7
24	122	14	7/32	0.24	5.050	17.0	5.381	19.1
27	112	14	7/32	0.27	4.851	14.0	5.271	16.0
30	104	14	7/32	0.30	4.903	12.8	5.421	14.8
33	100	14	7/32	0.37	4.732	11.1	5.359	13.2
36	118	12	7/32	0.36	4.630	10.0	5.376	12.2
39	111	12	7/32	0.39	4.564	9.1	5.439	11.5
42	106	12	7/32	0.42	4.541	8.5	5.556	11.0

Deflection Equation:

$$x = \frac{D_l k W r^3}{EI + .0614 E' r^3}$$

Where
x = horizontal deflection in inches;
D_l = deflection lag factor;
k = bedding constant (related to bedding angle);
W = vertical load on pipe in pounds/inch;
r = mean radius of pipe, in.;
EI = flexural rigidity of pipe wall (pound-square-inches);
E' = modulus of soil reaction;

Trench Load;
w = 110 lb/cu ft
$K\mu$ = 0.13
Bd = Outside Diameter Pipe + 24 in.
D_l = 1.0
k = .096

*The above table is intended only as a design example and is based on a limiting deflection of $D^2/4000$ in., where D is the inside nominal pipe diameter in inches.

†For correlation of E' values with degree of densification and soil types, reference is made to the paper *Modulus of Soil Reaction Values for Buried Flexible Pipe,* by Amster K. Howard, Journal of the Geotechnical Engineering Division, ASCE, Vol. 103, No. GT 1, Proceedings Paper 12700, January 1977, pp. 33-43.

soil density measurements made to verify that design assumptions are met.†

Bedding Details

Bedding details for both trench and embankment conditions are illustrated in Fig. 5.1 and 5.2 to represent typical construction practice in bedding concrete pressure pipe. Concrete cradles, illustrated in both figures, are not ordinarily used for semi-rigid pipe.

The type of bedding may be chosen after determining how much soil support must be provided to limit pipe deflection as required in AWWA C303. For trench installations, it is recommended that trench widths be kept as narrow as possible to minimize the external load, while still providing adequate working space.

Unstable Foundations

Situations are sometimes encountered where soil has insufficient supporting strength to provide a stable foundation for pipe. Examples of this would be water-saturated soils such as those found in many areas along coastlines or in swamps. In such cases, support can be provided in one of several ways:

1. The trench may be over-excavated and a more stable foundation such as gravel or shell may be placed to provide stability and adequate support beneath the pipe.
2. Piles may be employed to provide stability at each joint. A saddle type of support may be constructed on top of each pile. More detailed information on pile supports can be found in Chapter 8 of this manual.
3. A mat foundation may be used to distribute the pipe and over-burden load over a wider area. An example of this is a timber mat.

References

1. SPANGLER, M.G. *Soil Engineering.* International Textbook Company, 2nd ed. (1960).

Bibliography

Concrete Pipe Design Manual. American Concrete Pipe Association (1974).
Design and Construction of Sanitary and Storm Sewers. ASCE, Manual and Reports on Engineering Practice No. 37 (1968).

---

†A lag factor (D_1) of 1 is normally assigned to concrete pressure pipe. A higher factor may be warranted if lines are not to be pressurized for a long time. A discussion of the "re-rounding" effect of internal pressure is included in ASCE Preprint No. 1259, *Design of Buried Pressurized Flexible Pipe* (Jul. 13, 1970), by R.K. Watkins.

CHAPTER 6

Installation by Trenching or Tunneling — Methods and Equipment

Scope

This chapter covers the common practices and requirements for installation of concrete pipelines. Normal practices and standard equipment are discussed and brief descriptions of special equipment and methods that have been used in the installation of very large diameter pipe and tunnel liners are also given. The actual job requirements, of course, will depend upon several controlling factors which are discussed briefly in the following sections.

Specific details of trenching, laying, and backfilling can only be determined after consideration of the design and type of pipe to be used. The installation methods and equipment selected by the contractor should be based on the actual requirements and conditions of the project.

A general checklist at the end of this chapter lists subjects that require consideration when planning for the installation of pipeline. Only a few of the subjects may be pertinent on small jobs; a greater number become applicable for larger and more complicated jobs.

Trenching — General Considerations

The trenching operation is a major item in the overall planning of the project. Many potential problems for both the contractor and the owner can be avoided if there is adequate planning prior to excavation. It is necessary to recognize that the engineering properties of the soils to be excavated and to be used as backfill are very important. In the final stages of planning, the right-of-way width should be determined by considering the size of pipe, depth of burial, type of material, safe slopes for the trench, and any special requirements such as separation of top soil. The principles of soil mechanics, applied to both trenching and backfilling, will lead to a safer, better, and more economical pipeline installation.

Planning of trenching procedures. Pre-construction conferences between the contractor and the project engineer should include in-depth discussions of the contractor's proposed trenching, pipe laying, and backfilling operations to be sure that they are compatible with the contract requirements and specifications. Differences can be settled more easily before construction begins. A meeting with the pipe manufacturer at this time is of equal importance so that manufacturing and delivery can be coordinated with the installation schedule. Special pipe or fittings generally require more lead time to fabricate, and delivery schedules of these items should be agreed upon as soon as possible.

Prior to the start of any site work, the contractor should establish and lay out rights-of-way, work and storage areas, access roads, detours, protective fences or barricades, and pavement cuts. He must also consider provisions for adequate access to businesses and residential properties.

National Safety Act. Safety measures have always been important to protect both workmen and the public, and the subject of safety has gained increasing attention since the advent of the Occupational Safety and Health Act. A project can be interrupted, or even shut down, for what may seem to be a minor infraction of the safety act. It is imperative that responsibility for safety be assigned to an authoritative person who has full knowledge of the rules, regulations, and requirements of both OSHA and state and local agencies.

Construction survey. A thorough reconnaissance survey of the entire right-of-way should be made to eliminate potential problems during construction. The size and operating space requirements of equipment, the conditions of streets, the intensity of traffic, and the location of trees and overhead lines should be noted and considered. Photographs of the pertinent areas are useful and time-saving. Substructures and utilities must be located and planned for in advance. Public records and records of utility companies can supply this information.

Surveys to accurately establish the horizontal alignment and profile for the pipeline route, with adequate cross sections utilized where necessary, should be made on all portions of the route. When data from other sources are used, survey ties should be checked. Traverses should be closed for horizontal control, and benchmarks checked for vertical control. Concrete pressure pipelines are manufactured to fit the plan and profile. Variances between recorded and existing distances, either vertical or horizontal, can cause unnecessary difficulty and expense.

The transfer of line and grade to the excavation work from control points established by the engineer is the responsibility of the contractor. Care should be taken to preserve these references because, generally, a charge will be made for re-establishing grade stakes that are carelessly destroyed. The line and grade may be transferred to the bottom of the trench by the use of batter boards, tape and level, or patented tape- and plumb-bob units. Batter boards and batter-board supports must be suspended firmly across the trench and span the excavation without measurable deflection. Another method of setting grade is to set from offset stakes or offset batter boards and double string lines with the use of a grade rod that has a target near the top. When the pipe invert is on grade, a sighting between grade rod and two or more consecutive offset bars or the double string line should show perfect alignment. A third method, usually used on larger pipe and flatter grades, requires that the line and grade for each length of pipe be set by means of a transit and level from either on top of or inside the completed part of the pipeline. Another and newer method incorporates the use of laser beams.

Site preparation. The amount of site preparation required will vary — under some conditions, none will be required; under other conditions, preparation can be a major item of the project cost. In some cases it can be considerable even in rural or undeveloped areas.

Operations that should be classified as site preparation are: clearing of trees, brush, or other plant growth; removal of rocks or unsuitable soils; construction of access roads and detours; relocation of existing drainage facilities; and the protection or relocation of existing utilities. The extent and diversity of this list is obvious; however, it should be stressed that the success of keeping the project on schedule depends on thoroughness of planning and timely execution of the site preparation work.

Open-Trench Construction

The specific details of trenching, laying, backfilling, and the testing of any pipeline depend on several factors, among which are: the type and purpose of the line, its size, operating pressure, whether it is an urban, suburban, or rural location, and the type of terrain in which it is to be laid.

The details are also affected by the depth of the trench and, as previously mentioned, the type and classification of the native soil and its suitability for use as backfill material.

Trench dimensions. Because of load considerations (discussed in Chapter 4), the depth and width of the trench should be as shown on the plan and profile or as specified by the project engineer. If these dimensions are exceeded, improved bedding may be required. The width of the trench at and below the top of the pipe should be only as wide as necessary for the proper installation and backfill of the pipe. Good practice will not exceed the outside pipe diameter, plus 30 in., for larger sizes. The width of trench above the top of the pipe to the ground surface is related to its effect on adjoining facilities and surface conditions.

In open country, economic considerations often justify sloping the sides of the trench above the top of the pipe to the ground surface. This may eliminate substantial amounts of sheeting and bracing, unless safety regulations require it. In improved or paved streets, it may be desirable to restrict the trench width at the ground surface to reduce cutting of pavement and restoration costs.

Excavation. Regardless of the type of excavating equipment employed, the trench should be dug to approximate grade with sufficient width to permit proper placing of backfill bedding material around the pipe.

With favorable ground conditions, excavation can be accomplished in one operation; under more adverse conditions it may require several steps. The trench bottom should receive careful attention and adequate provisions for maintaining grade. Pipe foundations are discussed in more detail later in this chapter.

Excavating machines and methods. Trench-excavating equipment employed varies with soil conditions, trench depth, terrain, and the contractor's preference. The more common types are backhoes, draglines, clamshells, power shovels, front-end loaders, and rotary trenching machines.

Backhoes. Backhoes are used extensively for the excavation of trenches with widths exceeding 2 ft and a depth up to 25 ft. They can be used in relatively restricted areas and in nearly all types of soil. Trench widths associated with various backhoe bucket capacities are listed in Table 6.1. The backhoe is also used frequently with a cable sling for lowering pipe into the trench.

Draglines. In open country or in a wide right-of-way, a dragline can be used. In cases of very deep trench excavation, 30–50 ft, a dragline can be used for the upper part of the excavation and a backhoe operating at an intermediate level can relay material to the dragline for removal to the spoil bank or trucks at the surface.

Clamshells. A clamshell is best employed when soil conditions or underground

TABLE 6.1
Trench Widths and Backhoe Bucket Capacities

Bucket Capacity — cu yd	Minimum Trench Width — in.	
	With Side Cutters	Without Side Cutters
⅜	24–28	22
½	28–32	27
¾	28–38	28
1	34–44	34
1¼	37–46	37
1½	38–46	38
2	50–58	50

structures restrict the use of other types of machines. In wet or unstable soil that requires close sheeting, the use of vertical-lift equipment such as the clamshell allows the bracing and sheeting to be installed as required.

Front-end loader. In wide, deep trenches a front-end loader can be used as an auxiliary to a backhoe or dragline. The backhoe or dragline can be used to excavate the upper part of a trench, leaving the bottom bench for a front-end loader to place its spoil within reach of the backhoe or dragline.

Trenching machine. Trenching machines are used to excavate trenches of moderate depth and width in light, cohesive soils. Under such conditions a trenching machine can make rapid progress at relatively low cost.

Sheeting and bracing. The primary function of trench sheeting and bracing is to prevent a cave-in of the trench walls or areas adjacent to the trench. Responsibility

Backhoe, which has excavated trench, is also used to lower pipe into trench

Sluicing being utilized for additional compaction during backfilling of large diameter pipeline

for adequacy of any required sheeting and bracing is usually delegated to the contractor. If sheeting is to be removed, consideration must be given to additional loads that may be transferred to the pipe (see Chapter 4). The design of the system of supports should be based on sound engineering principles of soil mechanics and materials to be used. Design must comply with applicable safety requirements.

Skeleton sheeting. In trenches with relatively stable soil, skeleton sheeting consisting of pairs of planks set vertically with spreaders or with trench jacks may be used. Horizontal spacing will vary with soil conditions.

Continuous sheeting. Continuous sheeting involves the use of panels prefabricated with plank stiffeners. The panels are set vertically, with spreaders or trench jacks spaced as required. For wider and deeper trenches or for more unstable soil, a system of wales and cross struts of heavy timber is often used. The horizontal wales are used to distribute the pressure.

Trench shields. In some soil conditions it is economical and practical to use a prefabricated unit that is at least as long as one section of pipe. The units are called laying shields or trench shields and are pulled forward as the trenching and pipe-laying progresses. Although they do not support the trench walls, they do protect workers from sluffs and fall-ins. Trench shields should be used only where the trench width is not critical and the shield is high enough and of sufficient strength to protect the workers from caving soil.

Steel sheet piling. In non-cohesive soils combined with groundwater, it may be necessary to use continuous steel-sheet piling to prevent soil movement. Steel-sheet piling can be installed so that it is relatively watertight, and, if necessary, dewatering with trench-bottom sump pumps can be employed.

An example of skeleton sheeting

Special conditions. In limited reaches of pipelines it is often necessary to make special provisions for supporting or bracing a trench. Excavations beside existing structures may be subject to surcharge loads. In such cases the trench bracing or supports must be adequate to withstand these loads, not only to prevent cave-ins, but also to prevent any movement or possible damage to the structure itself.

Dewatering. If unexpected and not planned for, groundwater can be a serious problem during excavation and pipe laying. It is a condition that the contractor should be aware of prior to bidding the project.

The trench should be dewatered before pipe is laid and kept dry until the pipe has been installed with the prescribed bedding conditions and backfill placed to a height at least above the groundwater level.

Storm drains or adjacent water courses, when present, can be used to dispose of the water. In some cases, it may be possible to drain the water through the completed portion of the pipeline to a point of satisfactory discharge.

The trench may be over-excavated and backfilled to grade with crushed stone or gravel to facilitate drainage to the point of removal. An excessive amount of groundwater, particularly where it creates an unstable soil condition, may require the use of a well-point system consisting of a series of perforated pipes driven into the water-bearing strata and connected to a header pipe and pump.

If trenching is necessary in coarse water-bearing material, turbine well pumps can be used to lower the water table during construction.

Pipe foundations. Firm cohesive soils, if undisturbed, provide adequate pipe

Pipe installation taking place within a trench shield

foundations when properly prepared. The bottom of a trench should be accurately graded and bell holes dug to accommodate pipe joints with projecting bells. The pipe barrel should be in contact with the trench bottom for its full length. Although sometimes performed, careful trimming and shaping of the bottom to fit the pipe barrel is costly and unless a special shaping machine is used, it can be very difficult to accomplish the desired degree of accuracy. Over-excavation in depth, followed by backfilling to grade with selected material to provide uniform bedding for the pipe, is increasingly being used because it is both practical and economical. When a soft trench bottom is encountered this procedure may be necessary for stabilization. The stabilizing material must be thick enough to prevent movement of the soft subgrade up into the bedding material. The required depth of stabilizing material should be determined by tests and observation.

In cases where a trench bottom cannot be stabilized with a selected backfill and where intermittent areas of unequal settlement are anticipated, special foundations for the pipe may be necessary.

If the trench bottom is solid rock, it must be over-excavated to make room for a selected material that will support the pipe uniformly. The trench bottom should be cleaned of all loose or projecting rocks prior to placement of the bedding material.

The design and required supporting strength of the pipe will always be based on a load factor that has been determined for various bedding conditions. For this reason, it is imperative that the complete pipeline be installed with the actual bedding requirements as described in the contract specifications. Pipe beddings

Pipe trench with steel sheet piling

have been defined and classified for assignment of appropriate load factors in Chapter 5.

Pipe Installation — General

Proper and careful installation of concrete pressure pipe is of benefit to both the owner of the pipeline and the installing contractor. Acceptance of nearly all pressure pipelines must be based on a performance test of some kind.

Installation procedures for the several types of concrete pressure pipe, although basically similar, may have different requirements in order to attain successful installation. The pipe manufacturer should provide the installer with this information and instructions in the proper method of laying pipe.

Pipe delivery. Concrete pressure pipe is rarely a stock item. Thus, it is common for fabrication and installation to be carried on simultaneously. For this reason delivery schedules should be resolved by the manufacturer and the contractor. When changes are required, the pipe manufacturer should be notified as soon as possible so that his production schedule can be modified.

Job-site storage. Pipe can be stored or stocked at the job site prior to the start of laying. In storage areas, care should be taken to place the pipe so that it can be reached for movement to the trench with as little extra handling as possible. Pipe delivery can be scheduled to coincide with the installation, and normally the pipe is strung along the right-of-way in the proper sequence for laying.

Regardless of storage conditions, every precaution should be taken to prevent damage to the pipe. Pipe ends are particularly vulnerable to damage from impact or point loading resulting from contact with rocks or other obstacles on the ground.

Pipe handling. Pipe should, at all times, be handled with equipment designed to prevent damage to either the inside or outside surface of the pipe. When support stulls have been installed in the pipe by the manufacturer, they should remain until the pipe has been placed in the trench and backfilled.

Installing pipe on granular foundation

Laying the Pipe

Special pipe foundations for the trench bottom on which the pipe is to be laid will be described in the job specification. Normal installation is made with a firm foundation that is free of rocks or other objects and prepared true to line and grade. Uniform bearing for the full length of the pipe barrel should be provided.

Equipment. The equipment to be used in placing the pipe in the trench depends on the size, weight, and length of the pipe sections. The most versatile of the usual equipment is a crane that is capable of handling a wide range of pipe sizes and weights and which provides the operator with the best control over his load while jointing the pipe. Other types of equipment are (1) a backhoe equipped with a cable sling or strong-back for lowering pipe into the trench, and (2) the side-boom tractor for smaller, lighter pipe. For very large and heavy pipe sections, equipment specially designed for laying and jointing the pipe may be required.

Making the joint. Pipe should be supported free of the bedding or foundation during the jointing process to avoid disturbance of the subgrade. A suitable excavation should be made in the trench bottom to receive pipe with raised bells or to provide the necessary clearance required for grouting the exterior joint space. Any adjustments required to maintain grade should be made by scraping away or adding adequately compacted foundation material.

Both the bell and the spigot of the adjoining pipe should be clean and free of any dirt or mud. A thin layer of an approved commercial vegetable-type lubricant should be applied to the face of the bell, the gasket groove of the spigot, and the gasket. Generally, petroleum-based lubricants are not suitable for natural rubber gaskets. Rubber gaskets are designed and sized to provide the correct volume required to fill the groove and the annular space between the bell and spigot of a completed joint. This volume is based on a predetermined stretched diameter of the gasket and, for this reason, it is important that the stretch and the volume of the gasket be equalized around the entire circumference just prior to making the joint. This can be accomplished by inserting a rod under the gasket after it has been placed in the spigot groove and moving the rod rapidly around the full circumference.

"Pulling the pipe home." Because of the close tolerances in concrete pressure-pipe joints and because of the compression of the rubber gasket, considerable force is required to engage the joint. Come-alongs and power winches may be rigged in the pipe to provide this force. On smaller diameter pipe, a backhoe or other laying equipment *may* be used to force the pipe together. It should be emphasized, however, that the joint should be made with a straight-in pull. Raising the far end of the pipe so that the top of the spigot is inserted first and then lowering the pipe to insert the bottom half may result in a rolled gasket. Properly sized spacer blocks placed against the backside of the bell should allow workmen to easily maintain the correct joint space.

Checking the completed joint. The joint should be checked as soon as it is completed. A steel feeler gage approximately ½ in. wide and 0.015 in. thick can be inserted in the joint to determine, by feel, if the gasket is properly seated in the groove. To be assured of a leak-free joint, the gasket should be checked around the full circumference of the joint. If it is found to be out of place, the joint must be remade. On large pipe this type of checking is more easily done from the inside of the pipe; on smaller sizes it can be accomplished from the outside.

Interior joint protection. The exposed surfaces of steel joint rings may be protected by metallizing, by other approved coatings, or by pointing with cement–mortar. In some areas pipelines carrying potable water are not pointed with cement–mortar if the joint rings are protected with zinc metallizing or other protective coatings. When pointing of joint space with cement–mortar is used on pipe large enough to accommodate a workman, it is normally done with a hand trowel. The joint space should be clean before mortar is applied. The cement-to-sand ratio for the mortar mix should be 1:3, and the mortar consistency should be dry enough so that it will not fall down when placed in the top of the joint. Pointing of the inside joint space in small diameter pipe can be accomplished by "buttering" the back face of the bell with the mortar just before the spigot is inserted.

Grouting the exterior joint space. Grouting of the exterior joint space is accomplished by using a "diaper" or wrapper placed around the pipe and over the joint. The wrapper is held in place on either side of the joint with steel straps or bands. The ends are pulled together near the top of the pipe so that the access hole for pouring will allow the grout to be poured down one side and rise on the other. The grout must be wet enough to pour into the opening in the diaper near the top of the pipe and run all the way to the bottom. It should be rodded or puddled to ensure complete filling of the joint recess.

Backfilling

Good backfill procedures are important for the installation of all kinds of pipe. In the case of semi-rigid pipe, the load-carrying capabilities of the pipe can only be realized if the pipe is uniformly bedded or supported and the load uniformly distributed along the entire length. Rigid or heavy walled pipe, although not so dependent on good backfill consolidation for its load-carrying capabilities, must be uniformly supported along its bottom and under the haunches to prevent future settling or movement.

Backfill material. Backfill material must be free from rocks, tree stumps, broken pavement, or other unyielding solid objects. In the event that clean material is not available from the excavation, it should be imported. Unsuitable backfill material can result in voids around the pipe and subject it to irregular settling and the concentration of unusual forces for which it is not designed. Most job specifications indicate backfill material requirements; but in no case should solid objects or rocks that exceed 3 in. in diameter be used in the pipe zone. The best imported material is sand or a combination of sand and gravel.

Placing the backfill. Backfill can be placed with bulldozers or other equipment. Care should be taken to place the fill material uniformly on both sides of the pipe to prevent shifting or side movement of the pipe. When interior bracing or stulls have been placed by the pipe manufacturer they should not be removed until the backfill is complete. If the backfill is to be consolidated with water, the bracing should not be removed until the backfill has consolidated.

Backfill densification. When densification is required, there are two general methods in use: (1) "consolidation," which involves the use of water in a flooding or jetting operation, and (2) "compaction," which involves tamping with air tools, vibrating compactors, or other equipment. The method to be used depends on the kind of soil encountered.

If the soil is porous and self-draining like sand or gravel, then the best and fastest method is consolidation. In consolidation it is important for all of the material to be wet. If flooding is used, the fill should either be flooded in successive layers or be hosed continually as it is placed in the trench. If jetting is used, jet pipes should be inserted and worked up and down all the way to the trench bottom. Removal of jet pipes should be done slowly to prevent voids. If using the consolidation method, care should be taken not to float the pipe, thereby causing it to move off grade or alignment.

Cement-mortar grout being poured into diapered exterior joint space

When laying pipe in tight, non-draining soils like clay or other impervious soils, consolidation of the backfill should not be attempted. Water will not drain off and the backfill could take weeks to dry out enough to support the pipe properly. In these types of soils, densification should be accomplished by compaction.

Compacted backfill should be brought up in layers 6–12 in. in depth. During the compaction operation care should be taken to see that power tamping equipment does not damage the pipe.

Bedding the pipe. The most critical portion of the backfill is that which is placed under the haunches of the pipe (the lower third of the pipe's circumference). Time spent in being sure that this part of the backfill is well densified is worthwhile. The greater part of the earth load to be placed on the pipe will be concentrated in this area.

The fill from the haunches to the top of the pipe is also important, since it provides side support of the pipe. It should be brought up on both sides of the pipe evenly in layers so that the pipe will not be pushed out of line through unequal loading. This zone of backfill for semi-rigid pipe should extend to at least 6 in. above the top of the pipe before the remainder of the trench is filled. For rigid pipe the required compacted backfill zone will depend on the bedding angle used in the pipe design.

Backfill above the pipe zone up to finish grade should be densified according to the job requirements.

Surface Restoration

Upon completion of backfill the surface should be restored to a condition at least equal to that which existed prior to the start of construction. Concrete or asphalt pavements should be saw-cut and removed to a point beyond any caving or disturbance of the base materials. Before replacing permanent pavement, the subgrade must be restored and compacted until it is smooth and unyielding.

Tunnel Installations

Tunnel installations include various construction methods used to place underground pipelines that are made without a continuous open trench or disturbance of the ground surface. Concrete pressure pipe is also used as a liner in open-face mined tunnels and can either serve as a watertight conduit or be designed to provide needed structural support in unstable grounds.

Jacking methods. Jacking installations incorporate the use of large hydraulic jacks to push the pipe lengths through tunnels as the heading excavation progresses. An approach trench or jacking pit is excavated at the starting point to accommodate the backstop, jacks, pushing frame, and the jack pipe. The backstops and pushing frame are usually made from heavy timber; however, concrete anchor blocks are often used for large diameter pipe.

Jacking the pipe. Jacks made to operate in the horizontal position with capacities well in excess of the anticipated force required should be used. The lead pipe may be fitted with a steel cutting edge. The pipe should be cylindrical and have a smooth exterior surface.

When jacking, contractors have found it desirable to surround the outside of the pipe with a lubricant, such as bentonite, to reduce the frictional resistance. The pipe should be jacked upgrade if possible to facilitate drainage in the event that groundwater is encountered at the heading.

Once started, it is best if the jacking operation can be continuous, only interrupting to reset the jacks and, of course, to add additional pipe lengths. For extended delays, such as overnight, subsidence of the earth may result in an increased starting resistance difficult to overcome.

Many unique jacking methods have been devised over the years with installation accomplished at a relatively low cost. One such system, recently used to install 1600 ft of 36-in. diameter reinforced concrete pipe, 55 ft beneath an interstate highway in Minnesota, used a series of intermediate jacking rings that thrust 6-ft long sections of pipe forward independently of those ahead or behind. The pipe moved ahead like an earthworm, by a succession of expansions and contractions of the ring segments.*

Excavation. The excavation of the heading at the top and sides should be just slightly larger than the outside diameter of the pipe, but the bottom should be cut accurately to line and grade. In sizes less than 36-in. diameter, a mechanical auger or borer is generally used at the heading and the earth removed through the pipeline by conveyor. In large sizes with more working room, excavation can be accomplished either by hand or mechanical means. The material should be trimmed with care and the earth heading should not precede the lead pipe by more than is necessary.

It is always important to find out as much as possible about the soil along the pipe route, even to the point of taking bore samples. Large boulders or other unexpected obstructions are typical hazards when jacking pipe, and plans for overcoming them should be made before the operation begins.

Line and grade. In all jacking operations it is important that the direction of jacking be carefully established prior to the start of work. Guiderails for the pipe, set accurately for grade and alignment, must be installed in the approach trench or jacking pit. For large pipe it is desirable to have the guiderails set in concrete. Backstops must be strong enough and large enough to provide uniform support for the jacks so the pipe is not pushed out of alignment or grade.

Pipe joints. The joint design and configuration of most concrete pressure pipe is such that jacking forces can be satisfactorily transmitted from pipe to pipe; however, it is very important that protective joint spacers be used to prevent damage to the joints and to provide sufficient uniform bearing around the joint circumference. The pipe manufacturer should be consulted if the pipe is to be installed by jacking and his recommendations closely followed.

Tunneling methods. It is not within the scope of this section to attempt to cover or describe the various mining methods, machinery, or safety measures that are used in tunnel excavations. Federal and state regulations generally provide detailed construction and safety requirements, and the contractor should become aware of them and make every effort to comply. Although the primary purpose of these statutes is often to protect the workmen, they also help to avoid the catastrophic economic loss that may result from an accident.

The use of concrete pressure pipe as finish tunnel liners and methods of installing the pipe will be discussed in this section.

---

*Boring–Jacking System Moves Pipe Ahead Like an Earthworm. *Engineering News Record* (Mar. 27, 1975).

Primary liners. Large tunnels (5 ft or larger) in clay or granular material will normally require the use of a tunnel shield at the face. With a shield it is necessary to install a primary lining of sufficient strength to support the surrounding earth. The annular space behind the liner is filled with a cement grout as the tunneling and installation of primary lining progresses. Connections for pumping the grout should be provided in the primary liner.

Primary liners are usually made of steel or precast segmented blocks of reinforced concrete. Short sections of precast concrete oval rings are also used. They are of such dimensions that they can be moved through previously laid sections to the tunnel heading, and then joined with the last section that is placed.

Finish liners. When mined tunnels are required to be watertight or if they must operate under pressure, concrete pressure pipe that has proved to be an excellent and economical finish liner can satisfy both requirements. In rock tunnels not requiring a primary liner, the pipe may be installed as the finish liner. Grout connections can be built in the pipe wall for easy filling of the annular space between the pipe and the tunnel excavation. Pipe is also used as the finish liner in tunnels that have been mined with the primary liners placed as the excavation progresses. Grouting of the space between the pipe and the primary liner is accomplished the same as for rock tunnels and is discussed at the end of this chapter.

Installation. A variety of techniques has been used for transporting or moving concrete pipe into, or through, the excavated tunnel for placement. The methods vary from a simple skidding arrangement to special pipe-carrying machines that not only transport the pipe but are capable of positioning and joining a section of pipe to the one previously placed.

For short tunnels or smaller diameters, skids will normally be the most economical method for moving the pipe in the tunnel. The distance between the skid rails should not exceed 60 deg on the outside diameter of the pipe. If the rails are to remain permanently in place under the pipe, then they must be properly bedded and strong enough to provide uniform bearing for their full length. The area between rails should be shaped to clear the bottom of the pipe. It is advisable to keep both the skid rails and corresponding pipe surfaces well lubricated to minimize the sliding friction.

In long tunnels and for larger, heavier pipe, a system of rollers for carrying the pipe should be used.

Grouting. If grout is to be placed in the annular space between the pipe and the primary liner or the tunnel surface, it can best be accomplished by using pumping or pneumatic placing equipment suitable for handling the mixture to be placed. Positive provisions must be made to prevent flotation of the pipe. The grout mix must meet project specifications, but if it is to be pumped, the cement should not be less than 4 bags (376 lb) per cubic yard. To enhance its flow characteristics and facilitate placement, the mixture should contain about 2 percent of bentonite, by weight of cement, or other suitable plasticizer of recommended dosage.

A bulkhead for retaining the grout must be placed in the annular space at each end of the section that is to be grouted. At the start of the placing operation, the grout discharge pipeline should extend from the placing equipment to the bulkhead at the remote end. During placement, the grout discharge pipe must be positioned so that its discharge end is kept well buried in the grout at all times. After the grout is

Specialized equipment, called a "tunnomobile," being used to install large diameter pipe in tunnel

built up over the crown of the pipe, the grout placement line can be withdrawn as the grouting progresses. The placing of grout should continue under pressure until overflow in riser pipes of the required height is achieved. An alternate method of grouting is through ports installed in the pipe walls during manufacture. Sand is sometimes used in place of grout to fill the annular space between the pipe and the liner or the tunnel surface.

If the annular space between the pipe and the liner or tunnel surface is not to be filled with grout or sand, then flotation should be prevented by using blocks or holddown jacks.

Bibliography

Basic Water Works Management. American Concrete Pressure Pipe Association (1972).
Concrete Pipe Handbook. American Concrete Pipe Association (1965).
Design and Construction of Sanitary and Storm Sewers. ASCE, New York.
Design Manual, Concrete Pipe. American Concrete Pipe Association (1974).

Checklist
and
Guide for Installation Planning

Site Preparation
    1. Location and protection of existing structures and plantings.
    2. Existing utilities, protection, and access for use.
    3. Pavement removal.
    4. Storage locations — pipe and equipment.
    5. Rights-of-way.
    6. Construction of detours or traffic controls.
    7. Access roads.
    8. Barricades.

Excavation
    1. Selection of equipment.
    2. Classification of excavation (type of material).
    3. Limitations of trench width.
    4. Spoil placement.
    5. Excavation for appurtenances.
    6. Sheeting or bracing.
    7. Tunnels and/or jacking.
    8. Groundwater.
    9. Moving existing utilities.
    10. Installation of temporary utilities.

Installation and Backfill
    1. Pipe weight and length.
    2. Selection of equipment.
    3. Pipe jointing.
    4. Control of grade and alignment.
    5. Service connections.
    6. Connections to existing pipe.
    7. Preparation of trench bottom.
    8. Pipe bedding material — imported or native.
    9. Backfill placement and compaction.
    10. Backfill of appurtenances or structures.
    11. Pavement or sidewalk replacement.
    12. Removal of excess soil.
    13. Surface restoration.
    14. Acceptance tests.
    15. Measurement and payment.

Appurtenances and Miscellaneous Materials
    1. Air relief outlets and valves.
    2. Drain outlets, valves, and piping.
    3. Manhole outlets and structures.
    4. In-line valves.
    5. Structure construction.
    6. Encasement or cradles.
    7. Concrete and reinforcing steel.
    8. Cathodic protection systems.
    9. Measurement and payment.

CHAPTER 7
Thrust Restraining Methods

Unbalanced Forces
Pressure pipelines will experience unbalanced forces when flow changes direction (such as in elbows, wyes, tees, etc.), or cross-sectional area (such as in reducers), or is terminated (such as at bulkheads). These forces, if not adequately restrained, tend to disengage joints, as illustrated in Fig. 7.1. At a particular fitting, the unbalanced force is actually the sum of two forces: (1) static force due to internal pressure of the pipeline, and (2) dynamic force due to changing momentum of flowing water. Since most water lines operate at relatively low velocities, the dynamic force is usually ignored when computing thrust. For example, the dynamic force created by water flowing at 8 fps is less than the static force created by 1 psi.

Thrust Resistance
A fitting subjected to an unbalanced thrust has two inherent sources of resistance: (1) frictional drag from concrete dead weight of fitting, earth cover, and contained water, and (2) passive resistance of soil against the back of the fitting. If this type of resistance is not adequate to resist the thrust involved, then it must be supplemented either by increasing the supporting area on the bearing side of the fitting with a thrust block or by increasing frictional drag of the line by "tying" adjacent pipe to the fitting.

Thrust Blocks
Thrust blocks increase the ability of fittings to resist movement by increasing the bearing area. Typical thrust blocking of a horizontal bend (elbow) is shown in Fig. 7.2.

Magnitude of thrust force. Magnitude of the thrust force T is defined as follows:
$$T = 2\,PA\,\sin\frac{\Delta}{2}$$
where P = internal pressure (psi); A = the cross-sectional area of the pipe $\left(\frac{\pi D^2}{4}\right)$ using the *joint diameter* for D (in.); Δ = deflection angle of bend; and T = thrust force (pounds).

Calculation of size. Once thrust magnitude is determined, thrust block size can be calculated based on the bearing capacity of the soil:
$$\text{Area of block} = L \times D = \frac{T}{\sigma}$$
where $L \times D$ = area of bearing surface of thrust block (sq ft); T = thrust force (lb); and σ = safe bearing value for soil (lb/sq ft).

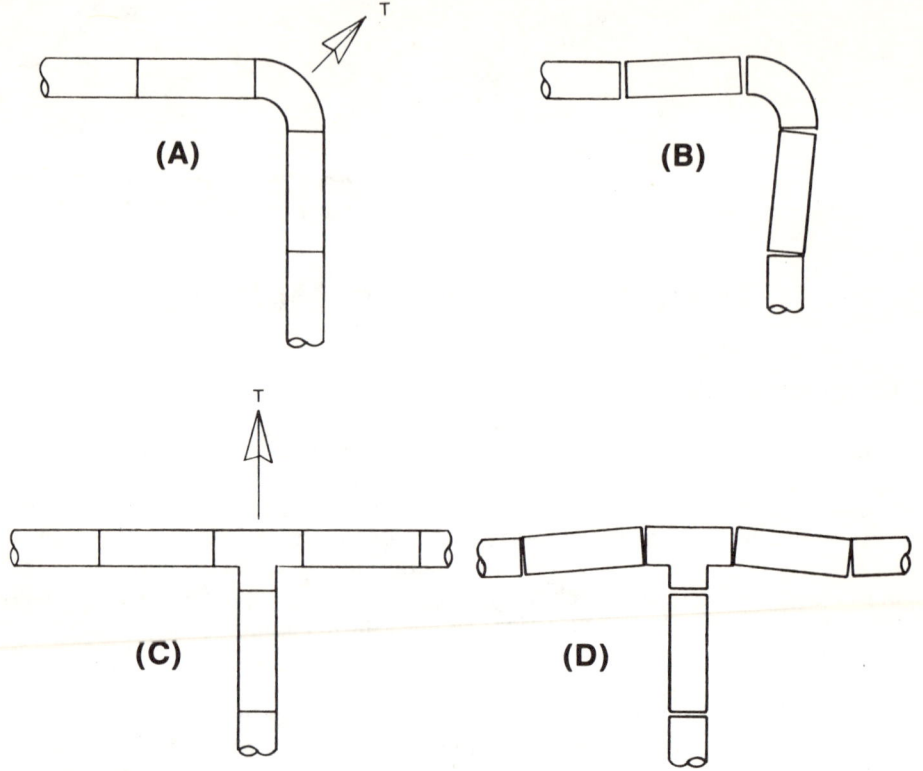

Fig. 7.1 Unbalanced thrusts and movement in pipeline
(A) and **(C)** Direction of unbalanced thrust is shown at elbow and tee.
(B) and **(D)** Mode of movement is shown at elbow and tee when unbalanced thrust is inadequately restrained.

Typical configurations. Determining the value for σ is the key to "sizing" a thrust block. Values can vary from less than 1000 lb/sq ft for very soft soils to several tons/sq ft for solid rock. Knowledge of local soil conditions is necessary for proper sizing of thrust blocks. Figure 7.2 shows several details for distributing thrust at a horizontal bend. Section A-A is the more common detail, but the other methods shown in the alternate sections may be necessary in weaker soils. Figure 7.3 shows typical thrust blocking of vertical bends. Design of the block for a bottom bend is the same as for a horizontal bend, but the block for a top bend must be sized to adequately resist the vertical component of thrust with dead weight of the block, bend, water in the bend, and overburden.

Proper construction essential. Even a correctly sized block can fail if it is not properly constructed. A block must be placed against undisturbed soil and the face of the block must be perpendicular to the direction of and centered on the line of action of the thrust *T*. A surprising number of thrust blocks fail due to inadequate design or improper installation. Many people involved in construction and design

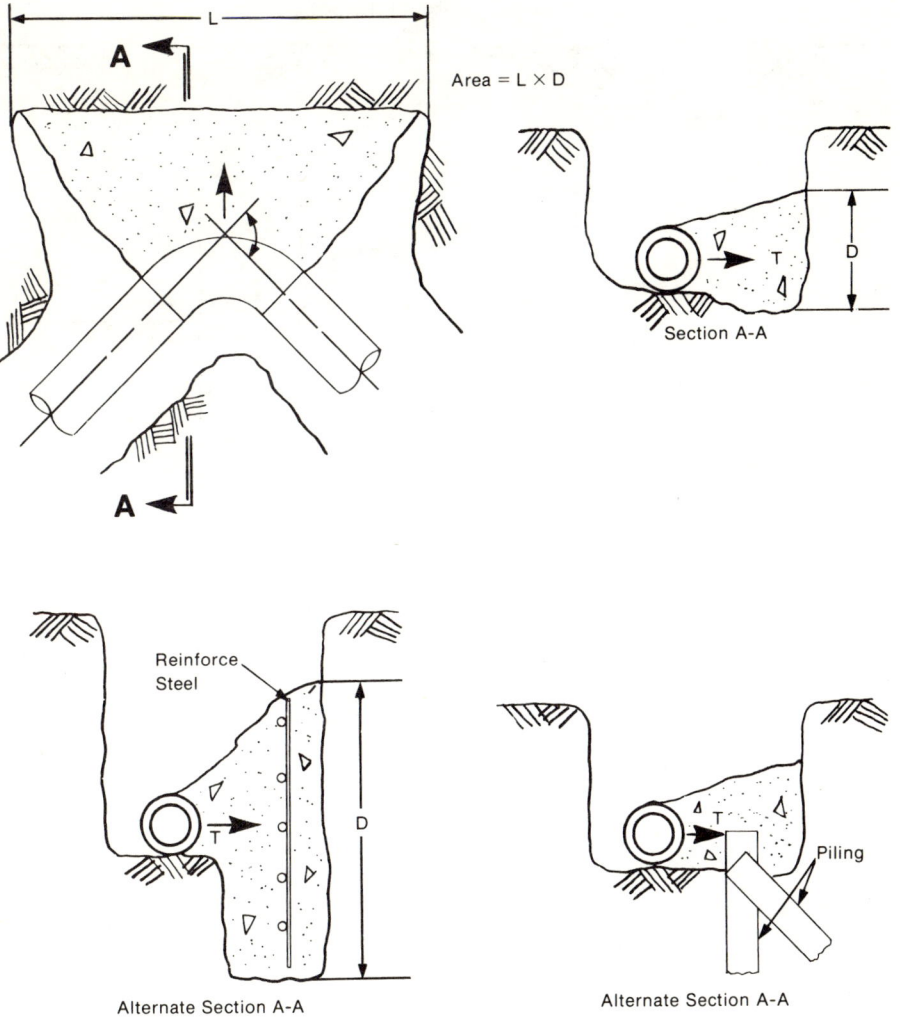

Fig. 7.2 Typical thrust blocking of a horizontal bend

do not realize the magnitude of thrusts involved. As an example, a thrust block behind a 36-in., 90-deg bend operating at 100 psi must resist a thrust force in excess of 150,000 lb. Another factor frequently overlooked is that thrust increases in proportion to the *square* of pipe diameter. A 36-in. pipe produces approximately four times the thrust produced by an 18-in. pipe.

Adjacent excavation. Even a properly designed and constructed thrust block can fail if soil behind it is disturbed. Thrust blocks of proper size have been poured

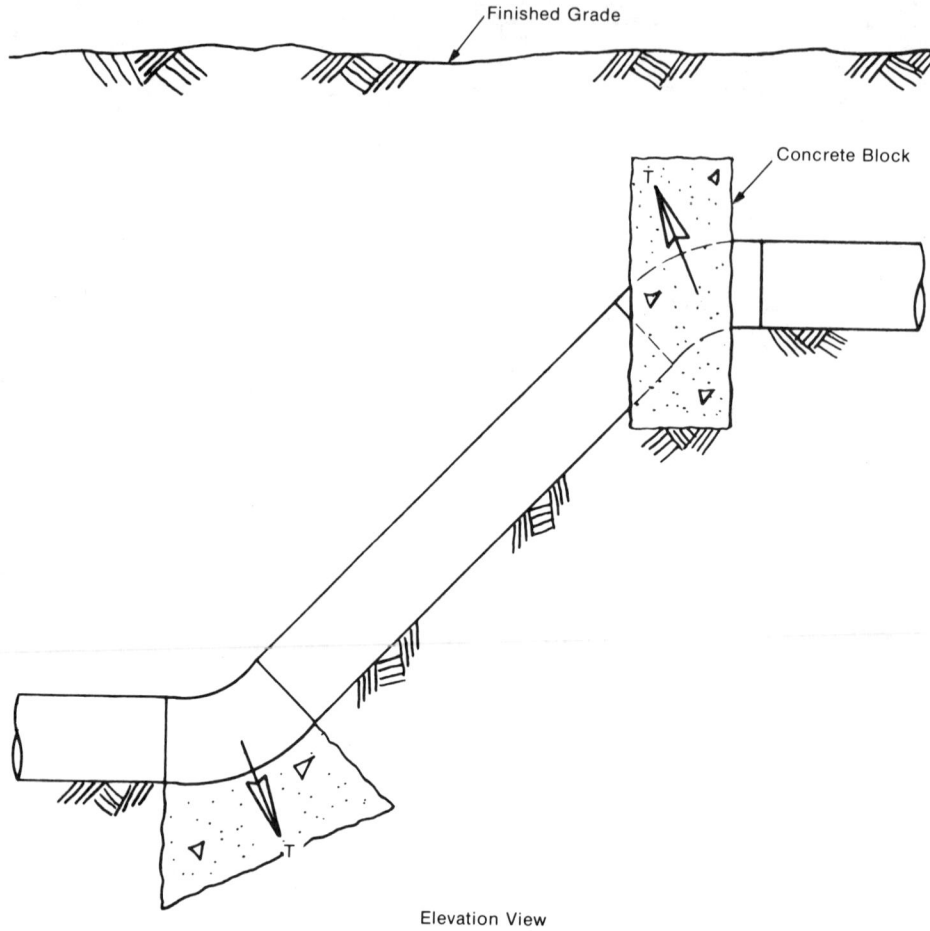

Fig. 7.3 Typical thrust blocking of vertical bend

against undisturbed soil only to fail because another utility or an excavation immediately behind the block collapsed when a line was pressurized. Awareness of problems that can occur with thrust blocks, adequate location records, and communication among various utility agencies is needed to prevent failures of thrust blocks.

Tied Joints

To avoid some uncertainties of thrust blocking, particularly the possibility of future excavations behind a block, many engineers choose to restrain thrust by the alternate method of increasing frictional drag of the line by tying adjacent pipe to the fitting. In order to determine frictional resistance of pipe against soil, the engineer must calculate or assume (1) values for weight of overburden, pipe, and water, and (2) a coefficient of friction for soil against pipe. Figure 7.4 shows a free-

Tee and reducer on large diameter line (note sandbags behind tee as forms for thrust block)

body diagram of the vertical forces used in determining frictional resistance.

Calculation of frictional resistance. As illustrated in Fig. 7.5, the frictional resistance F needed along each leg of the bend is $PA(1-\text{Cos}\Delta°)$. The component F_2 is balanced by the passive resistance of the soil. Frictional resistance of pipe against soil is equal to

$$f(2W_e + W_p + W_w)$$

where f = coefficient of friction between pipe and soil; W_e = overburden load (lb/lin ft); W_p = dead weight of pipe (lb/lin ft); W_w = dead weight of water in pipe (lb/lin ft); P = internal pressure (lbs/sq in.); and A = cross-sectional area of the pipe (sq in.).

Determination of overburden load used in calculations should be based on a backfill density and height of cover consistent with what can be expected when the line is pressurized. Values of soil density used in computing harness distance vary from 80 to 120 lb/cu ft, depending on degree of compaction. Selection of a value for coefficient of friction is dependent upon type of soil and roughness of pipe exterior. In identical soils, coefficient of friction on concrete pipe with mortar coating (C301 or C303 type pipe) will be greater than on pipe made of other materials that have smooth coatings. Values for coefficient of friction vary from 0.25 to 0.50. Tests conducted on buried pipe have shown that resistance to pipe movement approximately doubles after rainfall or if jetting of the trench consolidates the soil around the pipe. Length of pipe L to be tied to each leg of an elbow is calculated as

$$L = \frac{PA(1 - \text{Cos }\Delta°)}{f(2W_e + W_p + W_w)}$$

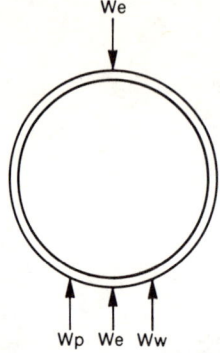

Fig. 7.4 External vertical forces acting on a buried pipe

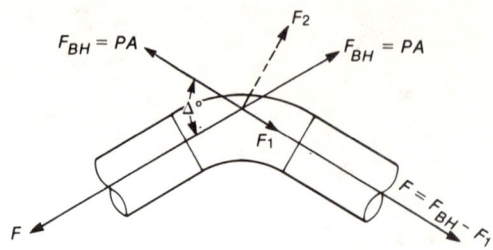

Fig. 7.5 Typical tied pipe force diagram

$F_1 = F_{BH} \cos \Delta° = PA \cos \Delta°$
$F_2 = F_{BH} \sin \Delta° = PA \sin \Delta°$
$F = PA (1 - \cos \Delta°)$

Transmission of tensile force through pipe. In addition to calculating pipe length to be tied to a fitting, designers must be sure that tied pipe lengths have sufficient continuous steel in the longitudinal direction to transmit thrust forces. Since thrust diminishes from maximum value at a fitting to zero at distance L from the fitting, requirements for longitudinal reinforcement can be calculated for the pipe length immediately adjacent to the fitting and pro-rated on a straight line basis for the rest of the pipe within the tied distance L (see following examples). In concrete cylinder pipe (C300, C301, or C303), the steel cylinder is used to transmit longitudinal forces, since it is continuous over full pipe length and is welded to the joint rings at each end. In noncylinder pipe (C302), thrust is transmitted through full length bars which are welded to the joint rings at each end of the pipe. The allowable stress in the steel cylinder or longitudinal bars should not exceed 12,500 – 13,500 psi at working pressure or 15,000 – 16,000 psi at test pressure.

Design examples. The following two examples will show the basic calculations for determining thrust, tied pipe lengths, and longitudinal reinforcing.

Example 1 *(See Fig. 7.6)*

Given:
- Pipe size — 30 in. I.D.
- Joint diameter — 34¼ in.
- Cylinder diameter — 33¾ in.
- Working pressure — 140 psi (test pressure 170 psi)
- $\Delta°$ — 75 deg
- Earth cover — 5 ft

Design assumptions:
- Soil density — 100 lb/cu ft
- Coefficient of friction — 0.3
- Pipe weight — 350 lb/lin-ft
- Axial stress in steel cylinder — 12,500 psi at working pressure; 16,000 psi at test pressure

THRUST RESTRAINING METHODS

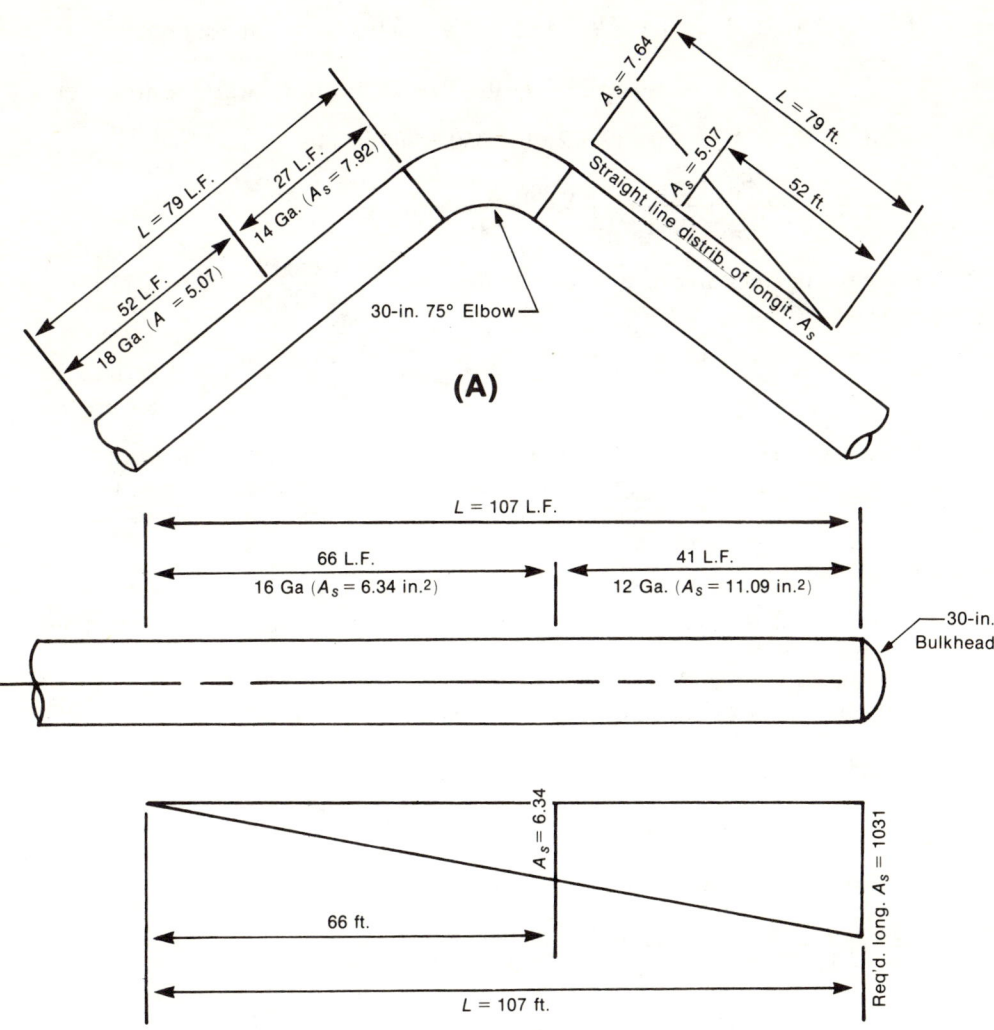

Fig. 7.6 Typical details for longitudinal reinforcement in tied pipe
(A) *30-in., 75-deg elbow at 140 psi with 5-ft earth cover. (170 psi test)*
(B) *30-in. bulkhead at 140 psi with 5-ft earth cover. (170 psi test)*

Preliminary calculations:
    Water weight   300 lb/lin-ft
    Earth load   2120 lb/lin-ft (from *Concrete Pipe Design Manual*)

    Cross-sectional area
        at joint   921 sq in.

$F = PA(1-\cos 75°) = 170 \times 921 \times (1-0.259) = 116{,}000$ lbs (at test pressure)

$ = 140 \times 921 \times (1-0.259) = 95{,}000$ lb (at working pressure)

$f(2W_e + W_p + W_w) = 0.3\,(2 \times 2120 + 350 + 300) = 1467$ lb/lin-ft

$$L = \frac{PA\,(1-\cos 75°)}{f(2W_e + W_p + W_w)} = \frac{116{,}000 \text{ lb}}{1467 \text{ lb/lin-ft}} = 79 \text{ L.F.}$$

Longitudinal reinforcement needed at elbow (at working pressure) $= \dfrac{F}{f_s} = \dfrac{95{,}500 \text{ lb}}{12{,}500 \text{ psi}} = 7.64$ sq in.

$ = \dfrac{116{,}000 \text{ lb}}{16{,}000 \text{ psi}} = 7.25$ sq in. (at test pressure)

Cross-sectional area of 33¾-in. O.D. steel cylinders: 18 gage = 5.07 sq in.
16 gage = 6.34 sq in.
14 gage = 7.92 sq in.

18 gage cutoff $= \dfrac{5.07 \text{ sq in.}}{7.64 \text{ sq in.}} \times 79 \text{ ft} = 52 \text{ ft}$

Use 52 ft of 18-gage cylinder tied pipe and 27 feet of 14-gage cylinder tied pipe. (See Fig. 7.6)

Example 2 *(See Fig. 7.6)*
  Given: Same as Example 1, except for bulkheaded condition.

$F = PA = 170$ psi $\times$ 921 sq in. $= 156{,}600$ lb (test pressure)

$ = 140$ psi $\times$ 921 sq in. $= 128{,}900$ lb (working pressure)

$f(2W_e + W_p + W_w) = 0.3\,(2 \times 2120 + 350 + 300) = 1467$ lb/lin-ft

$$L = \frac{PA}{f(2W_e + W_p + W_w)} = \frac{156{,}600 \text{ lb}}{1467 \text{ lb/lin-ft}} = 107 \text{ linear feet}$$

Longitudinal reinforcement required at bulkhead $= \dfrac{F}{f_s} = \dfrac{156{,}600 \text{ lb}}{16{,}000 \text{ psi}} = 9.79$ sq in.

$ = \dfrac{128{,}900 \text{ lb}}{12{,}500 \text{ psi}} = 10.31$ sq in.

Cross-sectional area of 33¾-in. O.D. steel cylinder: 18 gage = 5.07 sq in.
16 gage = 6.34 sq in.
14 gage = 7.92 sq in.
12 gage = 11.09 sq in.

16 gage cutoff $= \dfrac{6.34 \text{ sq in.}}{10.31 \text{ sq in.}} \times 107 \text{ ft} = 66 \text{ ft}$

Use 66 ft of 16-gage cylinder tied pipe and 41 ft of 12-gage cylinder tied pipe. (See Fig. 7.6)

54-in. cooling water line with six 20-in. riser outlets (harnessed joints used throughout for thrust restraint)

Practical variations. Both of these examples made use of only two different thicknesses of steel cylinder because of the relatively short distances involved. Another designer might have used other cylinder combinations, such as 14 gage and 16 gage in Example 1, and 12 gage and 18 gage in Example 2. Since various concrete pressure pipe manufacturers stock different sheet steel thicknesses, the common practice is to have the pipe manufacturer submit distances and cylinder thicknesses to be used subject to approval by the engineer.

Types of tied joints. In distance L, pipe joints must be "tied" so that thrust is transmitted across the joint. Generally, there are two types of tied joints: (1) welded and (2) harnessed.

Welded tied joints. Figure 7.7 shows a typical detail of a welded joint. Welding must be performed in the field in short sections of about 6-in. length to avoid overheating the gasket. The welder must "skip-weld" around the joint until the correct amount of weld is provided to transmit the thrust involved. Occasionally it is desirable to weld only a portion of the joint circumference. When analyzing such cases, it must be remembered that non-symmetrical weld patterns are eccentrically loaded, resulting in stress risers that subject the ends of the weld to a tearing type of failure. It is possible for a joint with a continuous 270-deg weld to resist approximately half as much thrust as a joint with the same total length of weld uniformly spaced around the full circumference. To avoid this problem and at the same time achieve welding economy, designers frequently specify that partial circumference welding be done in equal lengths centered at each springline.

Harnessed joints. The integrity of field-welded joints depends upon the ability and technique of the welder as well as trench conditions. Field-welded joints are expensive and time-consuming, which is a significant factor in congested areas where permissible length of open trench is limited. An alternate approach is to use harnessed (or restrained) joints, which provide a means of transmitting longitudinal thrust across the joints. Details of typical harnessed joints are shown in Fig. 7.8. Additional information on types of harnessed joints available and size and pressure limitations can be obtained from pipe manufacturers.

Combination Thrust Restraint System

Another type of thrust restraint system is a combination of restrained joints and thrust blocking. Figure 7.9 shows how an anchor pipe can be used to transmit thrust into undisturbed soil adjacent to the trench instead of behind the fitting. The anchor block must be poured against undisturbed soil and it may have to be reinforced. In certain locations this type of thrust restraint system can be used when a normal thrust block or complete harnessing system cannot be used.

Other Uses for Restrained Joints

Harnessed or restrained joints have other uses that are not related to thrust caused by internal pressure. If it is necessary to install a pipeline on a steep slope, it may be desirable to use harnessed joints or partially welded joints to keep the pipe from separating due to downhill sliding. Although the pipe may be capable of resisting downhill movement because of its own frictional resistance with the soil, the backfilling operation can sometimes provide enough additional downhill force to open the joint.

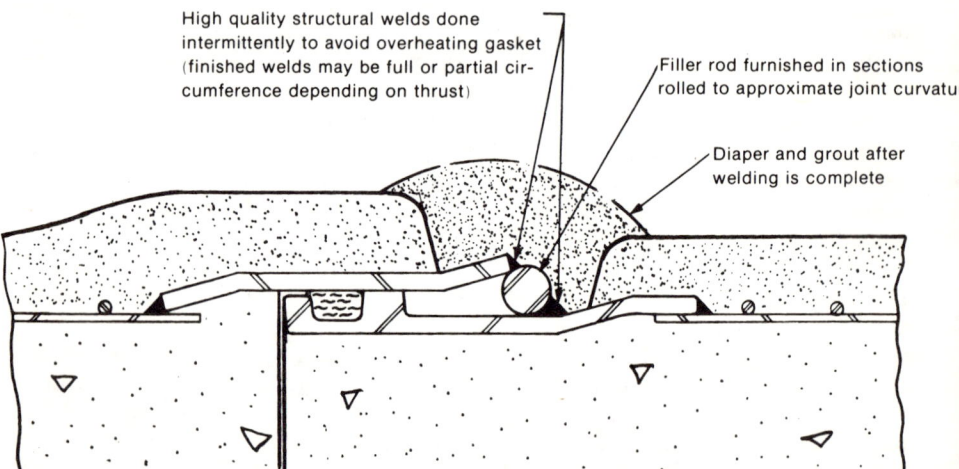

Fig. 7.7 Typical detail of welded tied joint

THRUST RESTRAINING METHODS

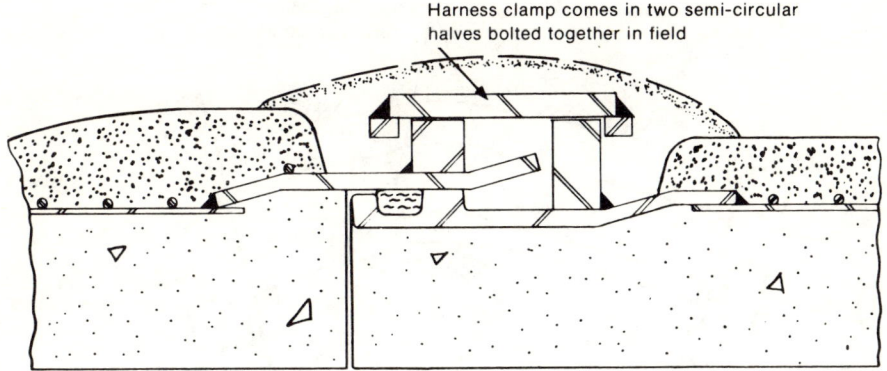

(A) Clamp type harness

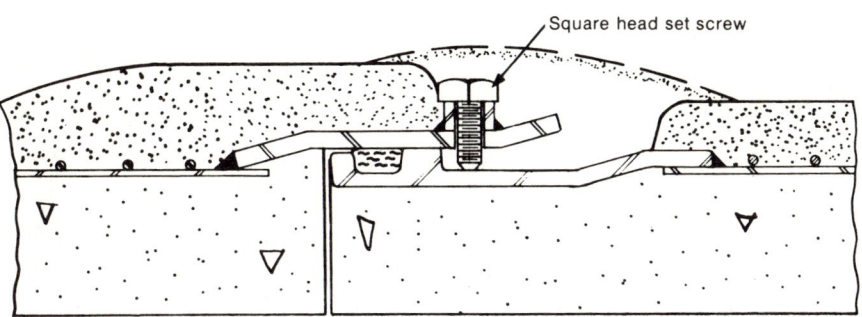

(B) Bell bolt type harness

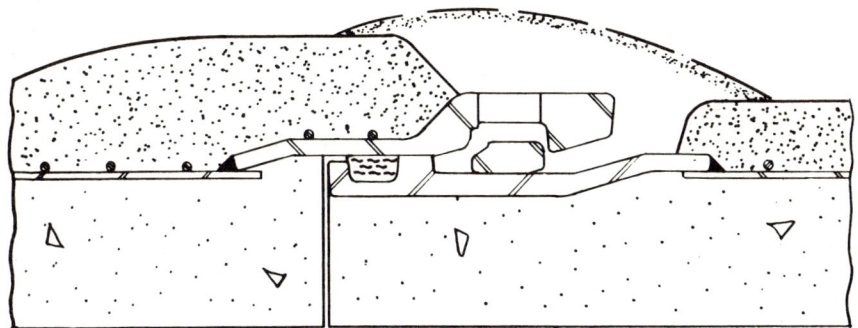

(C) Snap ring type harness

Fig. 7.8 Details of typical harnessed joints

86 CONCRETE PRESSURE PIPE

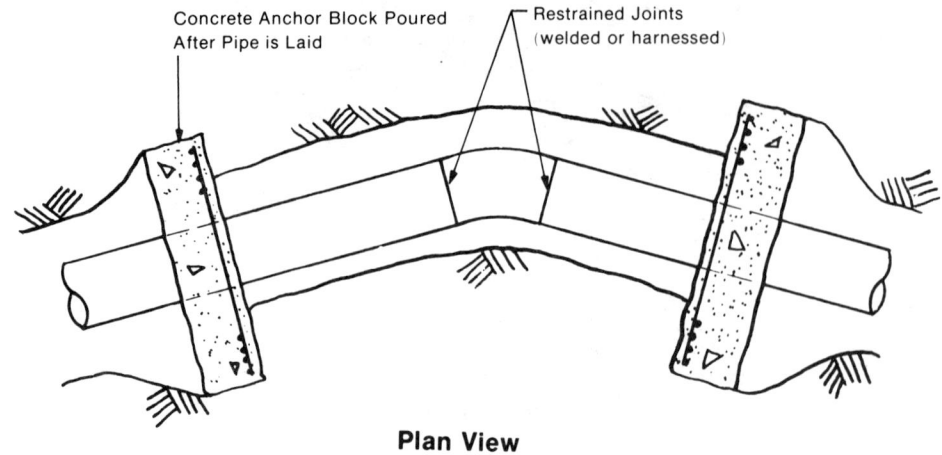

Plan View

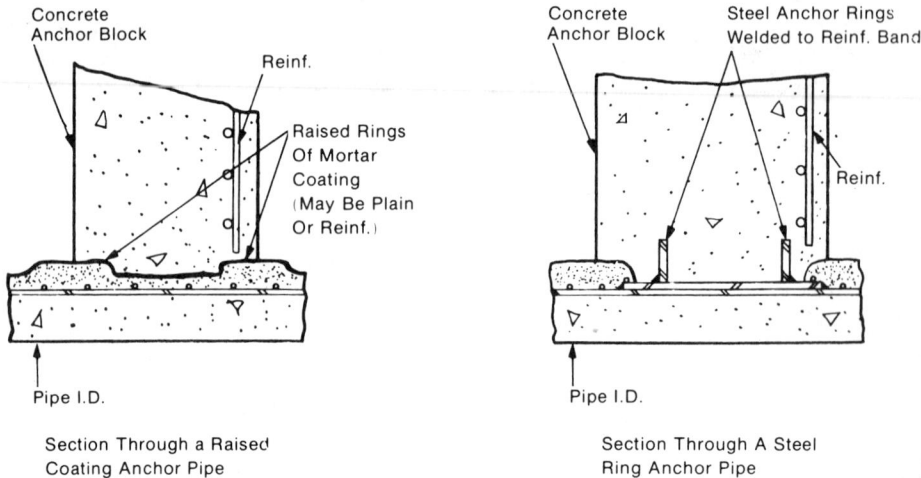

Fig. 7.9 Typical anchor pipe details

CHAPTER 8
Pile-Supported Installations

General

Although concrete pipe is normally bedded so that continuous and uniform support is provided, soil conditions or terrain that prevent this type of installation may be encountered along a pipeline. Pile-supported piers can be utilized to provide support for pipe that is laid in unstable soils or installed in aerial crossings. In either case, the pipe must be designed to span the supports as a beam and to resist the crushing load which is applied to the pipe at the pile support.

The standard rubber-gasketed joint is suitable for this type of installation. The joints may be placed over the piers, in which case the longitudinal bending moments may be computed on the basis of a simply supported beam; or they may be offset from the piers, with moments being computed on the basis of a continuous, pin-connected span.

The method of supporting pipe on a pier may vary from wood chocks to shaped saddles. The crushing load on the pipe will be related to the type of support provided.

Loads

Pile-supported pipe must be designed to resist the loads produced by both the pipe and the weight of water and, if buried, the weight of backfill over the pipe. The pipe is assumed to be completely unsupported between piers. The earth load on the pipe will depend on the soil conditions and installation procedures.

Buried pipe may also be subjected to live loadings, and if such is applicable to pile-supported pipe, the live load should be placed on the span so as to produce the *maximum* beam loading.

Beam Strength

The beam strength of pipe can be computed by considering the pipe as a reinforced concrete beam, circular in section, with the longitudinal steel in the pipe resisting the tensile stresses and the transformed area of concrete and steel resisting the compressive stresses. The moment of inertia of the section is computed in accordance with Appendix A, "Derivation of Moment of Inertia Equation." For design purposes the concrete compression value of $0.45\ f_c'$ is recommended, but under certain optimum conditions, values as high as $0.6\ f_c'$ have been used. Design stresses for steel tension of 15,000 psi for cylinder pipe and 10,000 psi for reinforced concrete pressure pipe are recommended. However, for installations that will not be subject to freeze–thaw cycles or other adverse environmental factors, a steel stress of 20,000 psi can be used for cylinder pipe. The foregoing steel stresses should be the result of the beam stress and any stress developed by longitudinal thrust.

Unstable soil conditions near the river bed required pier support for this 42-in. pipeline

Analysis of pipe as beam. If the pipe joints are placed directly over the support, the bending moments would be computed by the formula for a uniformly loaded simple span:

$$M = \frac{wl^2}{8}$$

where M = bending moment, in foot–pounds; w = weight of pipe, water, and external load, in pounds per linear foot; and l = span length, in feet.

In the preferred method of installation, the joints are offset from the supports as shown in Fig. 8.1. For this condition the maximum positive and negative moments will be*

At support $\quad M_1 = \frac{-wA}{2}(l-A)$

Between supports $\quad M_2 = \frac{w}{2}(l/2-A)^2$

where A = distance from support to joint, in feet.

Equating the two moments and solving for A to produce equal positive and negative moments:

$$A = 0.146\, l$$

and solving for M:

$$M = 0.062\, wl^2$$

---

*See Appendix B for derivation.

PILE-SUPPORTED INSTALLATIONS

50-ft spans of 42-in. concrete pressure pipe were used in this aerial line crossing of a highway and river levee

It will be noted that this moment is half of the simple span moment. Small deviations in dimension A can substantially change these bending moment relationships. Joint shear should be investigated where heavy earth loads are carried by pipe on pile. Joint shear is not a problem for aerial crossings.

The support system shown in Fig. 8.1 assumes balanced loads. To assure stability, end spans must be anchored by being buried in earth or by structural anchorage. Intermediate anchorages should be provided in long overhead crossings involving many multiple spans to preclude the domino effect which could result from imposed unbalanced loads.

Supports at Piers

The circumferential moments induced in the pipe wall at the pier supports can be

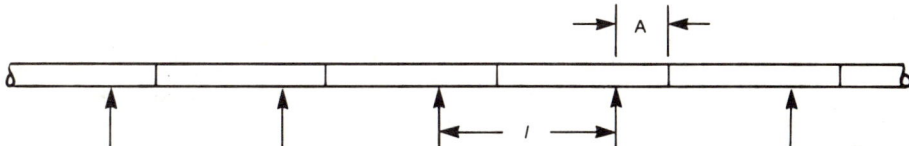

Fig 8.1 Configuration of pile-supported installations with joints offset from the supports

TABLE 8.1
Recommended Load Factors

Support Condition	Load Factor — L_f
90-deg concrete cradle	2.2
120-deg concrete cradle	3.0
Two point supports — 15° from vertical centerline	1.25
30° from vertical centerline	1.4
45° from vertical centerline	2.0

computed by the theories of J.H. Paris as given in "Coefficients for Large Horizontal Pipes." When the moments are related to those produced by a three-edge bearing load, load factors for various types of supports can be computed. Recommended load factors for typical supports are tabulated in Table 8.1.

Crushing load at piers. The pier reaction wl is imposed onto the pipe over the support length b as shown in Fig. 8.2. The unit load on the pipe over the support is

$$\frac{wl}{b}$$

It can be seen that designing the entire pipe length for this load unit is unnecessary and that the pipe adjacent to the support will resist the load along with that length which is directly over the support. One can define the total length of the adjacent sections together with the length of pipe over the support as the effective length $E.L.$ of the supporting section. The crushing strength for which the pipe must be designed is

$$Q = \frac{wl}{(E.L.) \times L_f}$$

Experience has shown that the effective length can be taken as the support length plus the I.D. of the pipe, in feet. Compressible bearing pads should be placed between the pipe and precast concrete saddle supports.

Aerial Crossings

Normally, concrete pipe is buried, which places it in a favorable environment by avoiding exposure to the extremes of weather conditions. If installation above ground is unavoidable, concrete pipe can be utilized when proper consideration is given to the effects of exposure.

It is most important that the flexibility of the joints be maintained to allow for expansion and contraction. This can be accomplished by filling the annular joint spaces with a mastic which will retain its flexibility.* In aggressive environments or climates subject to frequent freeze-thaw cycles, a second requirement for exposed pipe is the application of a seal coat paint over the concrete or mortar. The seal coat will prevent absorption of moisture by the exterior concrete and thus reduce potential damage from freezing and thawing. Seal coating will also prevent absorption by the coating of salts that may be present in the airborne moisture.

---

*Consideration must be given to proper anchorage of the pipe to the pipe support to resist forces which may result from temperature changes, hydrostatic thrust, and flotation.

PILE-SUPPORTED INSTALLATIONS 91

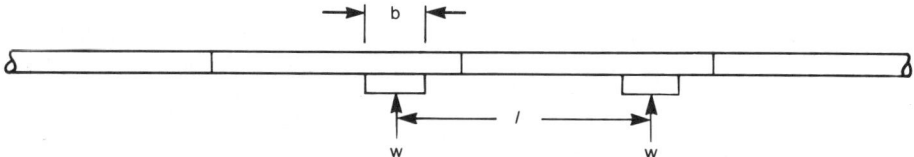

Fig. 8.2 Schematic support configuration

This 48-in. pipeline was installed on piers because of unstable soil conditions as well as to cross the stream

APPENDIX A

Derivation of Moment of Inertia Equation

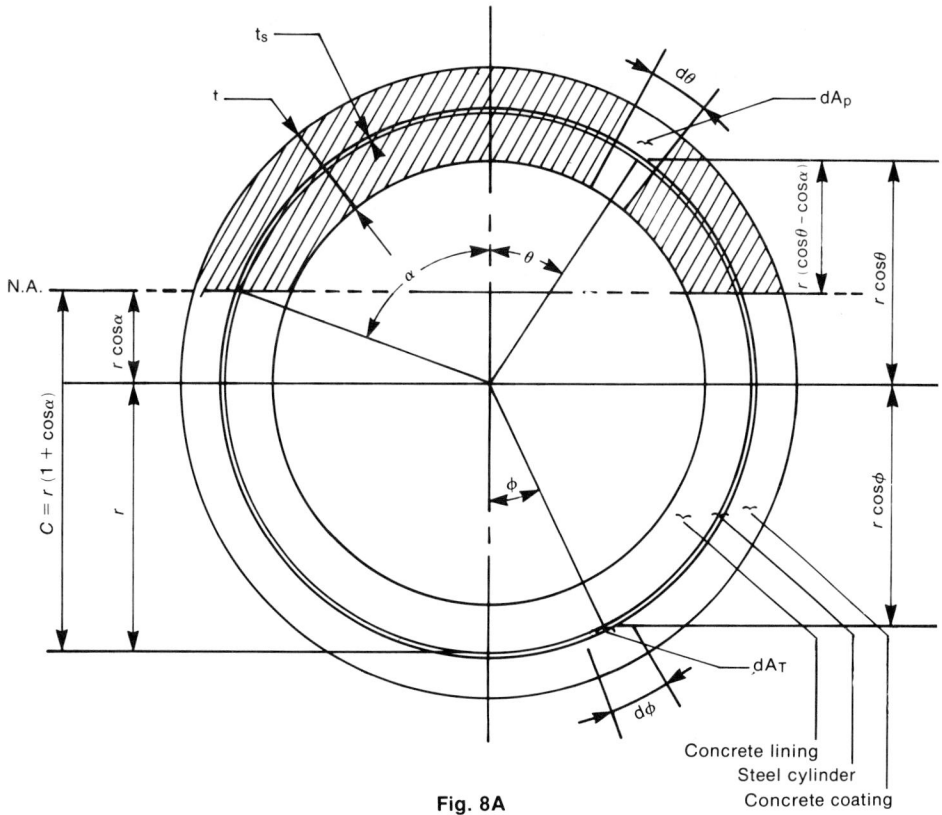

Fig. 8A

r = Outside radius of steel cylinder, in inches.
t = Total thickness of wall, in inches.
t_c = Thickness of concrete lining and coating, in inches.
t_s = Thickness of steel cylinder, in inches.

Area of concrete = $t_c r \, d\theta$
Area of steel = $t_s r \, d\theta$, or $t_s r \, d\phi$

$p = \dfrac{t_s}{t}$, or $t_s = pt$

and,
Area of steel = $ptr \, d\theta$, or $ptr \, d\phi$

$dA_p = t_c r \, d\theta + nptr \, d\theta$
$\quad\quad = tr \, d\theta + (n-1) \, ptr \, d\theta$
$\quad\quad = t \, [1 + (n-1)p] \, r \, d\theta$
$dA_T = nptr \, d\phi$

(1) Location of neutral axis.

$$2\int_0^\alpha r(\cos\theta - \cos\alpha)\, dA_P = 2\int_0^{\pi-\alpha} r(\cos\phi + \cos\alpha)\, dA_T$$

$$2\int_0^\alpha t[1+(n-1)p]\, r^2(\cos\theta - \cos\alpha)\, d\theta = 2\int_0^{\pi-\alpha} (npt)\, r^2(\cos\phi + \cos\alpha)\, d\phi$$

$[1+(n-1)p][\sin\alpha - \alpha\cos\alpha] = np[\sin\alpha + (\pi-\alpha)\cos\alpha]$
$(1-p)\sin\alpha = (\alpha - p\alpha + np\,\pi)\cos\alpha$

$$\boxed{\tan\alpha = \alpha + \frac{np\,\pi}{1-p}}$$

(2) Moment of inertia.

$$I = 2\int_0^\alpha r^2(\cos\theta - \cos\alpha)^2\, dA_P + 2\int_0^{\pi-\alpha} r^2(\cos\phi + \cos\alpha)^2\, dA_T$$

$$I = 2\int_0^\alpha t[1+(n-1)p]\, r^3(\cos\theta - \cos\alpha)^2\, d\theta$$
$$+ 2\int_0^{\pi-\alpha} nptr^3(\cos\phi + \cos\alpha)^2\, d\phi$$

Let $B = 2r^3 t[1+(n-1)p]$, and $C = 2npr^3 t$

$$I = B\int_0^\alpha (\cos\theta - \cos\alpha)^2\, d\theta + C\int_0^{\pi-\alpha}(\cos\phi + \cos\alpha)^2\, d\phi$$

$$I = B\left[\frac{\alpha}{2} + \frac{\sin 2\alpha}{4} - 2\sin\alpha\cos\alpha + \alpha\cos^2\alpha\right]$$
$$+ C\left[\frac{\pi-\alpha}{2} + \frac{\sin 2(\pi-\alpha)}{4} + 2\sin(\pi-\alpha)\cos\alpha\right.$$
$$\left.+ (\pi-\alpha)\cos^2\alpha\right]$$

$$I = 2r^3 t\left[1+(n-1)p\right]\left[\frac{\alpha}{2} + \frac{\sin 2\alpha}{4} - \sin 2\alpha + \alpha\cos^2\alpha\right]$$
$$+ 2npr^3 t\left[\frac{\pi-\alpha}{2} - \frac{\sin 2\alpha}{4} + \sin 2\alpha\right.$$
$$\left.+ (\pi-\alpha)\cos^2\alpha\right]$$

$$\boxed{I = 2r^3 t\left[(1-p)(\frac{\alpha}{2} + \alpha\cos^2\alpha - \frac{3}{4}\sin 2\alpha) + np\,\pi(\frac{1}{2} + \cos^2\alpha)\right], \text{(in.}^4)}$$

As the moment of inertia determined in the above equation uses the modular ratio n to convert steel areas to equivalent concrete areas, the moment of inertia is no longer a pure geometric property and caution must be exercised in its use. All steel stresses used with this moment of inertia to determine moment capacity of a pipe section must be in equivalent concrete terms. This condition also applies to the calculation of the moment capacity in the tensile zone, wherein the moment capacity must be calculated in terms of the allowable steel tensile stess divided by n.

(3) Distance between extreme tensile fiber of cylinder and neutral axis.

$\boxed{C = r(1+\cos\alpha), \text{ (in.)}}$

APPENDIX B

Continuous Span Beam Analysis

Joints are unharnessed and therefore may transmit shear but not moment.
$$R = wL$$
where w = weight per linear foot; R = reaction; and L = pipe length.

$$M_x = 0 = \frac{wL^2}{2} + PL - R(L-A)$$

$$= \frac{wL^2}{2} + PL - wL(L-A)$$

$$P = w(L-A) - \frac{wL}{2} = wL - wA - \frac{wL}{2}$$

$$P = \frac{wL}{2} - wA = w(\frac{L}{2} - A)$$

Maximum positive moment at y where shear = 0.
$$Bw = P$$

$$M_y = PB - \frac{wB^2}{2} = P(\frac{P}{w}) - \frac{w}{2}(\frac{P}{w})^2 = \frac{P^2}{2w}$$

$$M_y = \frac{w^2}{2w}(\frac{L}{2}-A)^2 = \frac{w}{2}(\frac{L}{2}-A)^2$$

Maximum negative moment at support Z.

$$M_Z = PA + \frac{wA^2}{2} = w(L/2 - A)A + \frac{wA^2}{2}$$

$$M_Z = \frac{wLA}{2} - wA^2 + \frac{wA^2}{2} = \frac{wA}{2}(L - A)$$

(continued on next page)

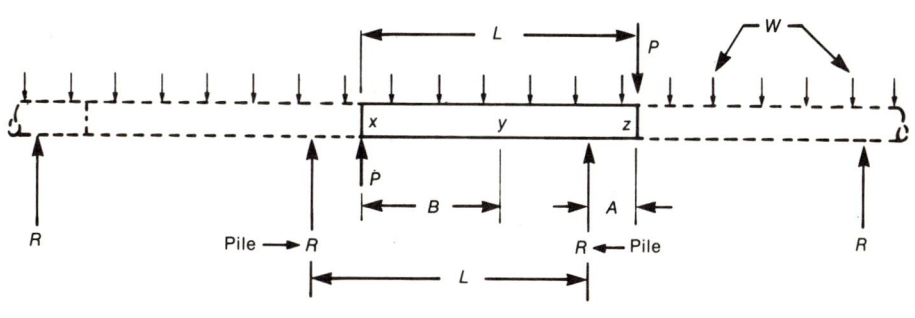

Fig. 8B

CONCRETE PRESSURE PIPE

Let $M_y = M_z$ and solve for A.

$$w/2 \, (L/2 - A)^2 = wA/2 \, (L - A)$$

$$\frac{L^2}{8} - \frac{LA}{2} + \frac{A^2}{2} = \frac{LA}{2} - \frac{A^2}{2}$$

$$A^2 - LA + \frac{L^2}{8} = 0$$

$$A = \frac{L \pm \sqrt{L^2 - \frac{4L^2}{8}}}{2} = \frac{L \pm \sqrt{0.5L^2}}{2}$$

$$A = 1/2 \, (L \pm 0.707L)$$
$$A = 0.146L$$

where

L — ft	A — ft
8	1.17
16	2.34
20	2.93
32	4.69
48	7.03

In order to allow construction tolerance for bents, design for $A \pm 1$ ft.

$$M_y = w/2 \, [L/2 - (A - 1)]^2$$
$$M_z = w/2 \, (A + 1) \, [L - (A + 1)]$$

Design for

Pile Spacing — ft	M
8	7.3 w
16	22.2 w
20	32.6 w
32	75.8 w
48	161.5 w

CHAPTER 9
Subaqueous Installations

Application

Subaqueous pipelines deserve special consideration both in design and installation. They can be located in both fresh water and seawater environments and cover a wide range of applications including
1. Crossing water or sewer lines under rivers, estuaries, or lakes.
2. Industrial cooling-water intake and discharge lines.
3. Sewer outfalls.
4. Intake lines for potable or irrigation water supplies.

Pipe Design Features

Evaluation of forces on subaqueous pipe. Earth loads on subaqueous pipe are generally of lesser magnitude than for ordinary installation. Unit weight of backfill is significantly reduced under water. Also, it is naturally easier to consolidate the bedding under water to obtain a more uniform pressure underneath and around the pipe.

54-in. diameter pipe, preassembled in 32-ft unit, being lowered with a double-bridle sling in a sewer outfall installation

Sections of pipeline that are installed in deep water which is not affected by waves, tidal action, or current are not subjected to significant external forces and should be installed with only sufficient cover to prevent bouyancy if that condition should occur. Sections of pipeline affected by waves, tidal action, and current are those that require the most critical treatment. Therefore, those sections must always be buried in trenches to prevent movement.

Hydrographic data of the proposed site should be obtained. A site should be selected where the shoreline is stable. In the design of adequate cover and protection to the pipeline as well as the appurtenant structures, consideration should be given to the problems of beach erosion, ships that drag anchor, wave action, and scouring and ice jams in rivers.

Rail-mounted gantry crane used for installing section of ocean outfall near shore

SUBAQUEOUS INSTALLATIONS

Buoyancy. The possibility of flotation of a pipeline must always be considered in the design of a subaqueous line. Buoyancy conditions should be considered with the pipeline empty, unless such conditions are impossible. The following observations are based on the experience of several consultants:

1. Pipelines with a bulk specific gravity of 0.5 or less will work their way through fine sand such as beach sand.
2. Pipelines with bulk densities between that of water and sand will have little or no tendency to ascend through fine sand.
3. If a pipeline is buoyant in water, then it is very likely to float out of sand or silt that is subject to wave action.
4. If a pipeline is buoyant in a liquified sediment, then agitation of the sediment will cause the pipe to rise. Cyclic changes in bottom pressure caused by storm waves can sometimes liquify slightly cohesive material and cause flotation. The same effect can also be caused by a variable water table, underground percolation, and unstable clay that is subjected to high water content.
5. If a pipeline is to float, then (1) the sediment must act like a fluid, and (2) the bulk specific gravity of the pipeline must be less than the specific gravity of the sediment.

Dead weight is generally the most practical method of anchoring offshore pipelines; coarse backfill material being most commonly used. Pipelines with heavy walls may also be used to overcome buoyancy. Table 9.1 shows the minimum pipe wall thicknesses needed to prevent flotation.

Subaqueous Pipe Details

Standard types of concrete pressure pipe can be used for subaqueous lines. Modifications can be made to the pipe to provide for joint-engaging assemblies, longer lengths, joint testing, and joint protection.

Joint-engaging assemblies. Subaqueous joint-engaging assemblies are available in many types and designs. Generally, they consist of heavy metal anchor sockets solidly cast into the wall at the ends of the pipe at each spring-line (see Fig. 9.1). Lugs are attached to the sockets to receive draw bolts that are used by divers to draw the bell and spigot together to make up the joint. These assemblies are of sufficient strength to pull the pipe home, but they cannot be relied upon to harness the joint against thrust. The bolting, which is temporary, does not require corrosion protection.

Pre-assembled lengths. To reduce the number of joints made under water, two or more lengths of pipe can be pre-assembled before laying. The length of pre-

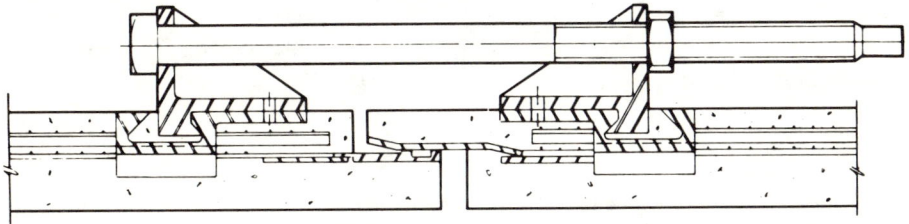

Fig. 9.1 Typical joint-engaging assembly

TABLE 9.1
*Minimum Wall Thickness for Neutral Buoyancy**

Nominal Diameter in.	In Seawater† in.	In Fresh Water‡ in.	Nominal Diameter in.	In Seawater† in.	In Fresh Water‡ in.
14	2¼	2¼	66	10¾	10¼
16	2¾	2½	72	11¾	11¼
18	3	3	78	12½	12¼
20	3¼	3¼	84	13½	13
24	4	3¾	90	14½	14
27	4½	4¼	96	15½	15
30	5	4¾	102	16½	15¾
33	5½	5¼	108	17½	16¾
36	6	5¾	114	18¼	17¾
39	6¼	6¼	120	19¼	18½
42	6¾	6½	126	20¼	19½
48	7¾	7½	132	21¼	20¼
54	8¾	8½	138	22¼	21½
60	9¾	9½	144	23¼	22¼

*Based on 150 lb/ft³ concrete.
†Approximate density 64 lb/ft³.
‡Approximate density 62.5 lb/ft³.

Five 84-in. diameter × 20-ft pipe sections were assembled on-deck into a 100-ft unit. The 100-ft unit was then mounted on precast concrete caps and cradles and installed as a single unit on piles. Two derricks lower strongback, cradles, caps, and 100-ft pipe unit

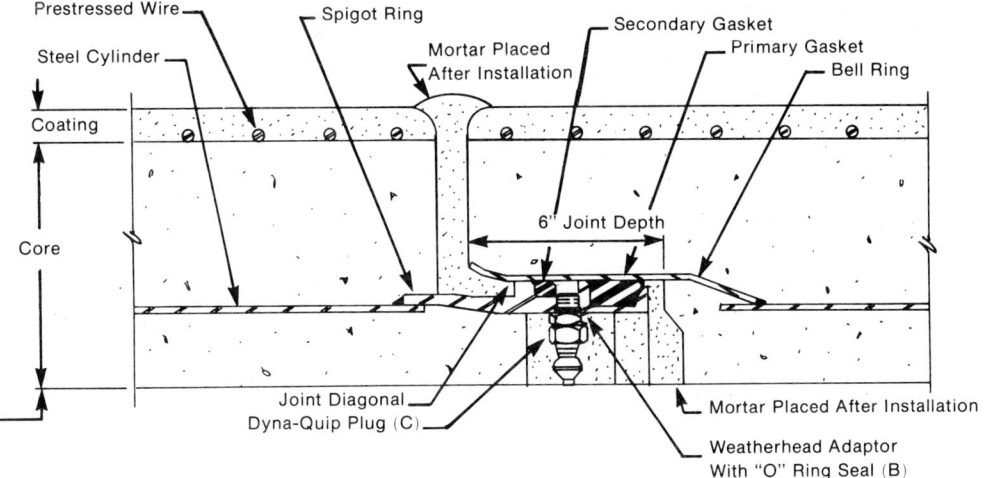

Fig. 9.2 Example of special joint design

assembled pipe is dependent upon available lifting equipment and the beam strength of the pipe. Pre-assembled lengths up to 200 ft of large diameter pipe have been laid successfully in deep water with proper equipment.

Joints. The conventional bell-and-spigot joint with rubber gasket is used for most subaqueous pipe joints. Special joint designs have been developed to allow testing each joint individually. A typical joint of this type is shown in Fig. 9.2. In some cases, the extra cost of such a joint may be justified.

Installation

The installation of subaqueous pipe requires special techniques and equipment depending on the size of the pipe, depth of water, and type of bottom.

Trenching. For soft, mucky material in shallow water, pile bents are preferred. There should be only one pile bent per pipe length situated behind the flexible joint (see Chapter 8). A pile-locating template may be required to position the pile bents. In solid material including rock, timber or concrete sills can be used. Concrete sills should have an embedded timber so that shimming wedges can be attached with spikes. Only one sill per pipe length is required. In place of sills in dredged trenches, burlap bags filled with lean dry-mix concrete or sand can be packed into the space between the pipe and the trench to form a support cradle.

Laying. A derrick scow or crane on a barge that is capable of handling pipe is needed for installation. In shallow water, a temporary causeway may be used. The pipe must be lifted with a double-bridle sling. Care must be taken to ensure that the harness lugs are on the horizontal centerline of the pipe. A spirit level laid across the ends of the pipe in a direct line with the lugs is generally used. This ensures against misalignment when two sections of pipe are connected with draw bolts. Before the gasket is placed in the annular groove on the spigot ring, it must be coated with a water-resistant vegetable soap. Surfaces of both the bell and the spigot must be similarly coated with the lubricant.

Large diameter pipe being lifted with a special double-bridle sling. One set of joint-engaging assemblies was installed at the ends of the pipe at the spring-line

Normally, with large diameter pipe, two divers must direct the positioning of the pipe under water so that the mouth of the bell is brought to the spigot of the pipe already in place. Then the pipe can be brought together by using the draw bolts. The nuts must be backed off a few turns to allow flexing of the joint. The pipe can be held in place with wood blocks or dry concrete bags to prevent movement during the backfilling operation.

Backfilling. Care should be taken to deposit the backfill uniformly on each side of the pipe. This is to ensure that the pipe will be supported evenly throughout its length and not be pushed off its alignment by the weight of the backfill. The backfill must be placed as soon as possible after installing the pipe. Coarse material should be used in areas subject to currents. In locations subject to ice forces, waves, and strong currents, armoring the backfill may be required. Rockfill, concrete bags, or concrete weights can be used for this purpose.

Testing

Subaqueous lines that carry water under pressure generally are tested by the application of pressure between valves or bulkheads in the same manner as with pipelines on land.

The working pressure of water intake lines is negligible, and leakage tests for these lines are often made by "infiltration" methods. A suction pump must be connected to a high point of each test section. Pipe cannot float if water is kept in the line. Infiltration is measured by the amount of water removed from the line in a prescribed period of time. If the pipe is heavy enough to prevent flotation, then the subaqueous installation may be tested by dewatering and conducting a visible inspection.

CHAPTER 10

Hydrostatic Testing and Disinfection of Mains

Hydrostatic Testing

Pressure pipelines are usually required to be hydrostatically tested by the contractor as a condition of final acceptance by the owner. With extremely long lines it is sometimes more convenient to test short sections as they are completed. Hydrostatic testing provides proof of pipe system integrity.

Test pressures. To test the system, the hydrostatic pressure should be at least equal to the maximum operating pressure of the line. Requiring a test pressure considerably in excess of the operating pressure serves only to increase costs since the pipe strength and size of thrust blocks or number of harnessed joints must be significantly increased. Test pressures are commonly specified as some value slightly greater than the operating pressure, such as 120 percent of operating pressure.

Testing after backfilling. Pipelines are generally backfilled prior to testing because of the limited amount of open trench space permitted for safety reasons. When harnessing is used as the method of thrust restraint, the harnessed sections must be backfilled prior to the test in order to develop necessary soil friction.

Apparent leakage. During the pressure test, the contractor is generally required to meter the amount of water that is added to the line to maintain test pressure. If the quantity added falls below a predetermined value referred to as "leakage allowance," then the line is considered acceptable. The term "leakage allowance" is a misnomer. It is intended only to give the contractor some allowance for trapped air, absorption of water by the pipe walls, take-up of restraints, and temperature variations during testing. Because of these factors, water may have to be added to maintain test pressure even if the line is not leaking. A more appropriate term for this quantity of water might be "apparent leakage." Most specifications require that observed leaks be repaired regardless of results of "leakage" measurements through metering equipment.

Preparation of line. Proper preparation prior to hydrostatic testing can help to keep apparent leakage to a minimum. Air valves should be properly located and checked to be sure they are operational. Lines with several high and low points should have small taps and corporation stops to bleed air while the line is being filled. All outlets should be plugged prior to testing.

Pretest soaking. The line should be filled at a slow rate to prevent air entrapment and should be left with a low pressure for 24 hr prior to testing. This will saturate the concrete lining and reduce the apparent leakage attributable to absorption by the pipe walls. Before testing, connections and equipment should be checked to see that they are in satisfactory condition.

Leakage allowances and test period. Leakage allowances are generally specified in the range of 10–50 gallons per inch of diameter per mile of pipe per 24 hours. This assigned value is intended only to give the contractor some allowance for apparent leakage, since any observed leaks must be repaired. A one- or two-hour test is generally ample to permit inspection of the right-of-way for evidence of leaks.

Bulkheads and thrust restraint. Bulkheads for use in conducting a hydrostatic test are available from pipe manufacturers. Generally, they have two outlets for filling and draining the pipeline and for bleeding air from the line. A system of thrust restraint is needed at the bulkheads. Most manufacturers furnish either a flat type bulkhead designed to be braced against a thrust block or a dished head type designed to be harnessed to the end of the pipeline. When bracing a flat type bulkhead against a thrust block, special care will be needed if timber is used to avoid a crushing type failure in the wood. The thrust block must be of adequate size and must be cast against undisturbed soil (see Chapter 7). For a harnessed type bulkhead there must be enough pipe harnessed to hold the thrust (see Chapter 7).

Safety precautions. A newly laid line is generally filled for the pressure test by connecting to an existing line. When this is not feasible, water can be pumped from a nearby source. The pump should be adequate to fill the line in a reasonable time. It will usually be different than the pump used to conduct the pressure test. Pumps should always be monitored while in operation to avoid accidental overpressuring of the pipeline. Positive displacement pumps should always have pressure relief valves in the system. Centrifugal pumps with a shut-off head within the limitations of the pipeline are preferred to piston pumps.

Two gages are desirable to provide a means of ensuring a correct pressure reading. Valves should be located at the air-bleed outlet and between the pump and the bulkhead. Meters for measuring leakage and pressure are generally furnished and calibrated by the owner.

Disinfection of Mains

After a line has passed the hydrostatic test, it must be disinfected if it is to carry potable water. Acceptable methods for disinfecting water mains are described in AWWA C601.[1] During construction of a water main, care should be taken to avoid unnecessary contamination of the pipeline's interior. The amount of contamination will affect the number of repetitions needed to attain acceptable disinfection.

References

1. AWWA C601. *AWWA Standard for Disinfecting Water Mains.* American Water Works Association, Denver, CO (1968).

CHAPTER 11

Corrosion Preventive Properties of Cement-Mortar Coating

History

The corrosion-preventive properties of cement–mortar or concrete coatings and linings* are well known.[1,2] These materials have been used since the early 1800's to prevent corrosion of pipe that is exposed to water and soil environments. Experience has shown that cement–mortar protection of ferrous pipe surfaces is still maintained after more than 100 years of continuous service.

Cement–mortar linings were used in 1836 to prevent tuberculation of cast-iron water pipe in France.[3] A method of cement–mortar lining of metallic pipe was patented in the US by Jonathan Ball in 1843, and a wrought-iron water line was installed with cement–mortar lining in Jersey City, New Jersey in 1845.[4] In 1855 a mortar-lined and coated steel pipeline was constructed for the city of St. John, New Brunswick, Canada. By the turn of the century, cement–mortar coatings and linings were used in many cities in the US to protect buried metallic water pipe from corrosion.

Protective Properties

The protective properties of cement–mortar cannot be explained in terms normally associated with most organic pipe coatings. Cement–mortar protects ferrous pipe elements (wire, cylinder, rebar, etc.) principally by passivation, whereas organic coatings attempt to insulate them from the environment. Cement–mortar coatings and linings for concrete pressure pipe are relatively thick and strong and contribute substantially to the structural strength of pipe, providing resistance to physical damage. Organic coatings and linings do not contribute significantly to the pipe strength.

Cement–mortar coatings and linings are unique because they provide ferrous metal surfaces with a passivating alkaline environment that is not adversely affected by moisture and oxygen.[5] In the absence of acids and high concentrations of chlorides or sulfides, embedded ferrous metals do not corrode and can even tolerate current discharge for a limited period.[6] The latter condition is not recommended.

Passivation. The unique corrosion-inhibiting properties of cement–mortar are mostly due to the chemical nature of portland cement. All hydrated cement products are chemically basic, and the pH of wet cement–mortar is about 12.5. This high rate of alkalinity is created primarily by calcium hydroxide which is produced from free lime in the cement reacting with moisture in the mortar. In the alkaline

---

*For this chapter, cement–mortar and concrete coatings or linings are synonymous.

environment, ferrous metal surfaces quickly develop a passivating iron oxide film before the mortar hardens. This protective film is assured by completely encasing the ferrous metal in quality mortar that is dense and of sufficient thickness to resist excessive penetration of oxygen, moisture, and high concentrations of dissolved corrosive salts. Small cracks in the mortar do not harm the protective film in most natural soil and water environments. As long as the integrity of the mortar cover is maintained, corrosion of the embedded steel will be prevented.

Impressed potentials. Another characteristic of mortar-coated ferrous metal is its resistance to corrosion under impressed potentials, or stray currents. The passive oxide film on the ferrous metal acts as a high resistance barrier to D–C current. Also, the highly alkaline cement–mortar surrounding the metal provides an abundance of hydroxyl ions to satisfy electrolysis reactions without iron being corroded. Any D–C current discharged from the ferrous metal initially consumes alkalinity in the cement; since the mortar is highly buffered, current discharge can be tolerated for some periods without damage to the metal.

Special Corrosion Control Measures

Like all pipeline protective materials, mortar coatings and linings must be properly designed and applied to perform satisfactorily. For proper design, the environment, the service requirements, and life expectancy of the pipeline must be known. For most soil or water environments, present-day design and application techniques are sufficient to ensure corrosion-free performance.

Exposed metal components such as outlets and other connections should be covered with cement–mortar after laying. Consideration should be given to electrically insulating connecting bare metal lines. For aggressive environments, special corrosion control measures, including the use of richer mixes, thicker mortar cover, dielectric top coatings, monitoring, and cathodic protection should be considered.[7,8]

References

1. SCOTT, G.N. Corrosion Protection Properties of Portland Cement Concrete. *Jour. AWWA,* 57:8:1038 (Aug. 1965).
2. MAYNE, J.E.O. & MENTER, J.W. The Mechanism of Inhibition of the Corrosion of Iron in Sodium Hydroxide. *Jour. Chem. Soc.* (Jan. 1954).
3. SPELLER, F.N. *Corrosion Causes and Prevention.* McGraw-Hill Book Co., New York (3rd ed., 1951).
4. *Steel Pipe Design and Installation.* AWWA Manual M11, American Water Works Association, Denver, CO (1964).
5. HAUSMANN, D.A. Steel Corrosion in Concrete, *Materials Protection,* 6:11:19 (Nov. 1967).
6. HAUSMANN, D.A. Electrochemical Behavior of Steel in Concrete. *Jour. American Concrete Inst.,* 61:2:171 (Feb. 1964).
7. ROBINSON, R.C. Design of Reinforced Concrete Structures for Corrosive Environments. *Materials Protection and Performance,* 11:3:15 (Mar. 1972).
8. Guide for the Protection of Concrete Against Chemical Attack by Means of Coatings and Other Corrosion Resistant Materials. *Jour. American Concrete Inst.,* 63:12:1305 (Dec. 1966).

CHAPTER 12
Tapping Concrete Pressure Pipe

Occasionally, it is necessary to provide an additional outlet on an existing concrete pressure pipeline. Tapping into concrete pressure pipe has become a routine procedure. Some pipe manufacturers have their own tapping services but there are also specialized service companies in the field. The procedures for tapping concrete pressure pipe while under pressure are basically the same as for other kinds of pipe. It is important for the designer to be familiar with the basic features of a pressure tap along with the particular requirements for concrete pressure pipe.

Threaded Pressure Taps up to 2 in. in Diameter

Mueller, A.P. Smith, or similar type drilling machines are normally used for making small taps in concrete pressure pipe. Carbide tipped masonry drills are recommended for best results.

Two types of saddles. Two general types of tapping saddles are used in tapping lined cylinder pipe. These differ basically in the manner in which they are secured to the pipe. The first type uses straps around the pipe (Fig. 12.1); with the second type the saddle is attached to the reinforcing wires or bars (Fig. 12.2). The second type is limited to a maximum of 1¼-in. taps.

Securing gland to cylinder. In general, the procedure is as follows:

Type 1 (Fig. 12.1)
1. Chip away mortar coating in an area slightly larger than the gland to expose reinforcing wires and cylinder.
2. Install saddle and straps.
3. Cut away wires carefully in order not to damage pipe cylinder. (For pressure taps in this size range, other types of tapping saddles are available which require cutting the reinforcing wires prior to installation of the saddle.)
4. Install gland gasket, gland, and gland nut. Tighten gland nut to attain seal between cylinder and gland gasket, taking care not to dent the cylinder or crack the core by over-tightening.

Type 2 (Fig. 12.2)
1. Chip away mortar in an area larger than saddle plate to expose reinforcing wires or rods and the cylinder.
2. Attach saddle plate to reinforcing wires with split studs and then anchor the wires that are to be cut to the saddle plate with anchor blocks.
3. Cut away wires carefully in order not to damage the pipe cylinder.
4. Install gland and gland gasket tightening against cylinder by using split studs.

Completion of tap. The following steps are common to both types of saddles described:
1. Tighten corporation stop into gland and open fully. Drill through cylinder and concrete core. Retract bit completely, close corporation stop, and remove tapping drill.
2. Coat all metal parts with a thin grout mixture, then build up a minimum of 1-in. thickness using a stiff mortar. As an alternate, straps can be protected by using a joint diaper and poured grout. The tap is now complete.

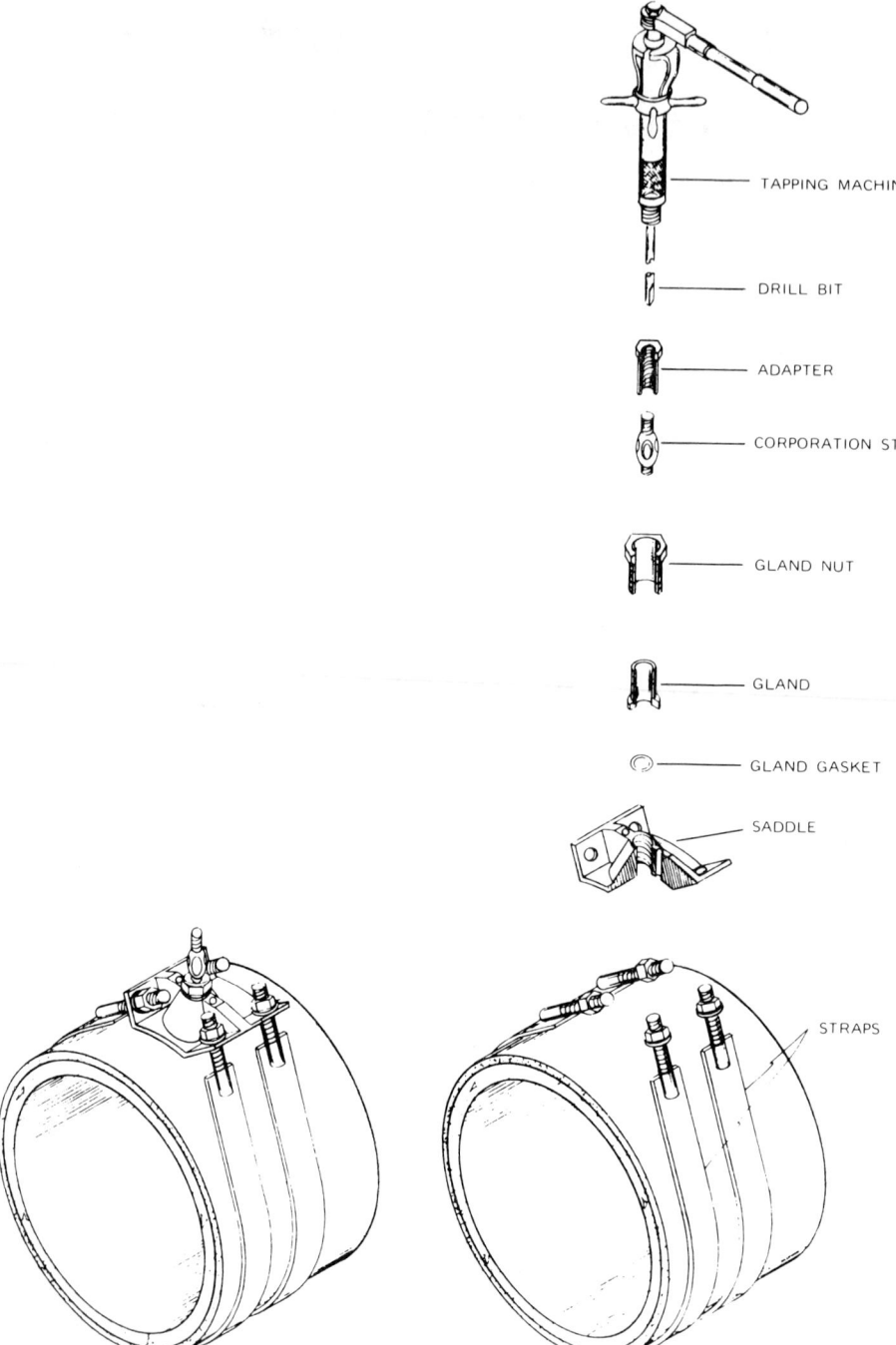

Fig. 12.1 Small-diameter pressure tap using saddle with straps

Pressure Taps for Flanged Outlets 4 in. and Larger

The following tapping procedure is a general method only. More specific information is available from pipe manufacturers. These taps are limited to the next size smaller than the pipe being tapped. Carbide tipped shell cutters, pilot drills, and power-operated automatic feed-tapping machines are recommended.

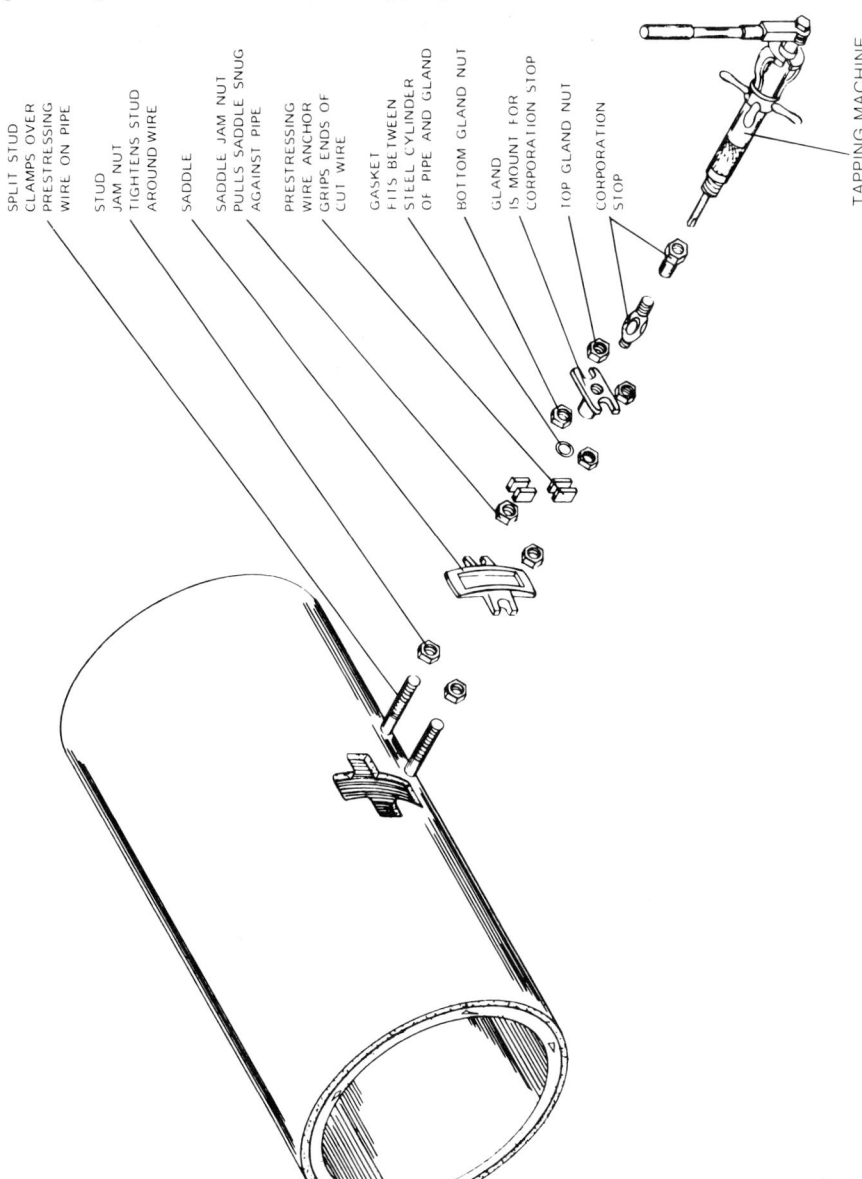

Fig. 12.2 Small-diameter pressure tap using split studs

Installation of a small-diameter threaded pressure tap using a saddle with straps

Tapping procedure. Numbered references in the following list are to details shown in Fig. 12.3.
1. Chip away mortar coating from the area (1) where the tap is to be made.
2. Fasten grout gaskets (3) around edge of saddle and position it in place, tightening all the straps (8) around pipe. Pour grout into prepared openings (4) in the saddle, filling the space between saddle and pipe. After grout has set, cut wires (5) to provide clearance for the gland to seal against the cylinder. For embedded cylinder pipe, the outer portion of the concrete core must be removed to expose the cylinder.
3. Position the gland (7) in the saddle (2) with its rubber gasket (6) against the steel cylinder. Tighten the outer circle of bolts connecting the two flanges compressing the gasket for a permanent water-seal. Place a blind flange on the gland flange and test the cavity for water tightness.
4. Remove gland flange and, for taps larger than 12 in. in diameter, secure concrete core being cut to steel coupon using special toggle equipment.
5. Fill the space between saddle and gland with grout and allow to set.
6. Attach a standard tapping valve to the gland flange with the inner circle of bolts.
7. Connect tapping machine to the tapping valve. Be certain that the saddle and gland will not have to carry the weight of the valve and tapping machine by providing support for both throughout and after the tapping procedure.

TAPPING

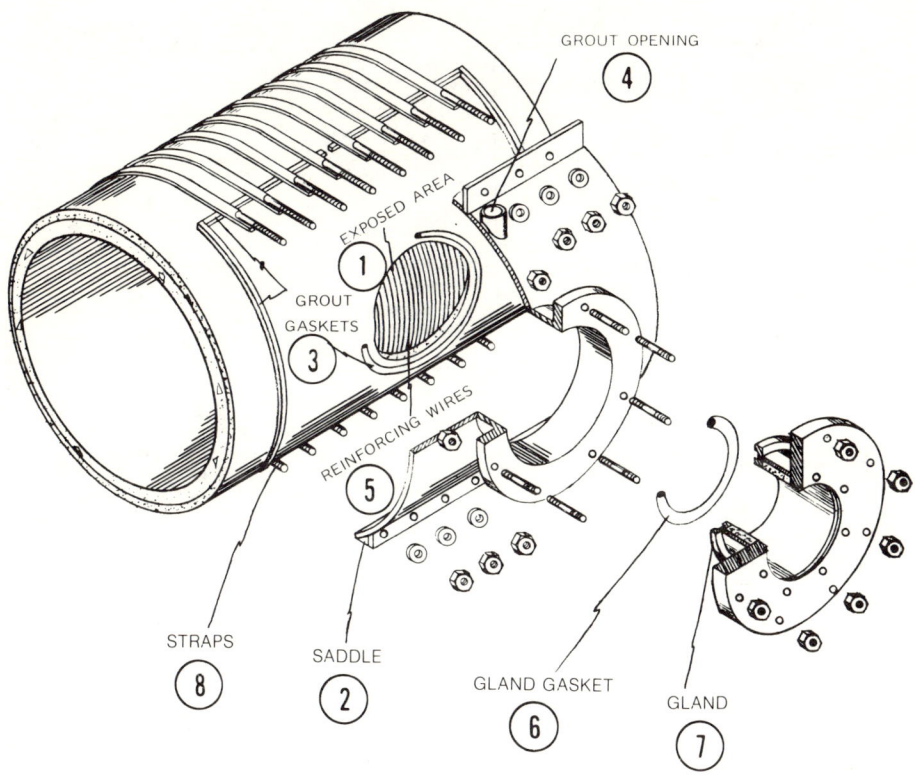

Fig. 12.3 Large-diameter tapping assembly

8. Advance the cutter by means of the handscrew through the opened valve to the steel cylinder of the pipe. When power is applied, the pilot drill will begin to cut the cylinder. Resistance to feed will suddenly increase as shell cutter contacts the pipe cylinder and begins its circular cut. When the feed screw has advanced the proper predetermined distance, the operator should know that the cut is completed.

Withdraw the cutting head past the gate and close the valve. Disconnect the tapping machine. Open the valve slightly to flush out any small cuttings that remain. Apply a protective coating of cement–mortar over the entire assembly.

Tapping Reinforced Concrete Pressure Pipe

Reinforced concrete pressure pipe (AWWA C302) can also be tapped with slight modifications to the previously described procedure.

Supplemental Data

Information concerning minimum excavation dimensions and range of outlet diameters for flanged pressure taps is given in Fig. 12.4.

112   CONCRETE PRESSURE PIPE

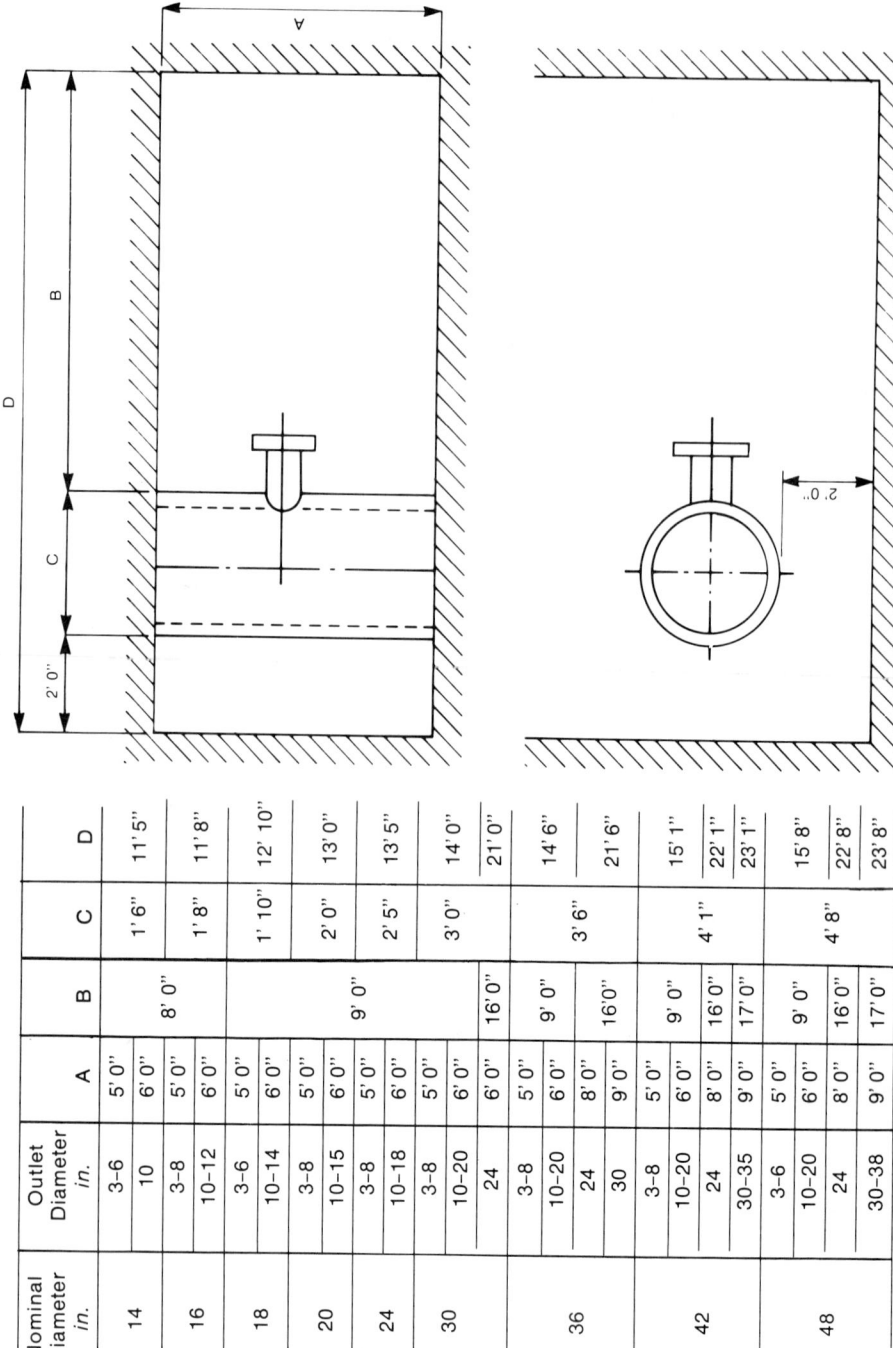

Nominal Diameter in.	Outlet Diameter in.	A	B	C	D
14	3-6	5' 0"	8' 0"	1' 6"	11' 5"
	10	6' 0"			
16	3-8	5' 0"	8' 0"	1' 8"	11' 8"
	10-12	6' 0"			
18	3-6	5' 0"	9' 0"	1' 10"	12' 10"
	10-14	6' 0"			
20	3-8	5' 0"	9' 0"	2' 0"	13' 0"
	10-15	6' 0"			
24	3-8	5' 0"	9' 0"	2' 5"	13' 5"
	10-18	6' 0"			
30	3-8	5' 0"	9' 0"	3' 0"	14' 0"
	10-20	6' 0"			
	24	6' 0"	16' 0"		21' 0"
36	3-8	5' 0"	9' 0"	3' 6"	14' 6"
	10-20	6' 0"			
	24	8' 0"	16' 0"		21' 6"
	30	9' 0"	17' 0"		
42	3-8	5' 0"	9' 0"	4' 1"	15' 1"
	10-20	6' 0"			
	24	8' 0"	16' 0"		22' 1"
	30-35	9' 0"	17' 0"		23' 1"
48	3-6	5' 0"	9' 0"	4' 8"	15' 8"
	10-20	6' 0"			
	24	8' 0"	16' 0"		22' 8"
	30-38	9' 0"	17' 0"		23' 8"

Fig. 12.4   Outlet sizes and excavation dimensions for flanged pressure taps

TAPPING

Pressure tapping of a 12-in. diameter flanged outlet from a larger-diameter pipe

Pressure tap of a 12-in. diameter flanged outlet from a 24-in. diameter pipe

CHAPTER 13

Guide Specifications for Purchase of Pipe

General

Concrete pressure pipe is sometimes procured directly by the owner for installation under a separate contract or for installation by agency personnel, or the purchase and installation may be included in a single contract. In either case, the specifications for the procurement of pipe can be simplified by making reference to the applicable AWWA standards and furnishing the supplemental details and information required by each standard in the section, "Plans and Data to be Furnished by Purchaser."

Specifications for pipe are usually included in the detailed specifications section of the project specifications. The following guide specifications may be helpful to the purchaser.

Pipe

General. The pipe manufacturer shall furnish all pipe, including fittings and special pipe, in the sizes and classes shown on the plans and in the bid schedule. Pipe shall conform to the requirements of AWWA C300, C301, C302, C303 and to the requirements of the specifications in this manual.

Design. Pipe shall be designed for the pressures and earth cover shown on the plans and the bid schedule for a minimum earth cover of feet in accordance with the methods outlined in the design appendix of the applicable AWWA standard. Typical trench sections with the required trench widths and pipe zone bedding are shown on the plans.

Submittals by manufacturer. Pipe design details and layout schedule shall be prepared by the pipe manufacturer and submitted in duplicate for approval in accordance with the appropriate AWWA standard.

Fittings and special pipe. Fittings and special pipe shall be fabricated in accordance with the details shown on the drawings. Welds that have been previously tested during the hydrostatic testing of cylinders need not be retested after fabrication of the fitting or special pipe. All other welds shall, at the option of the manufacturer, be tested by the air–soap method or the dye-penetrant method.

Marking of pipe. The manufacturer shall plainly mark the pressure class for which the pipe is designed on the inside of the spigot end of each pipe. In addition, all fittings and special pipe shall be marked with a number or station corresponding to that shown on the layout schedule.

Pipe from inventory. Pipe may be furnished from inventory if it meets the requirements of these specifications.

Certification. The manufacturer shall furnish a certification that all pipe supplied under this contract complies with the requirements of these specifications.

Rate of delivery. This section is applicable to "purchase" agreements only. When the specifications include both purchasing and the installation of pipe, the delivery requirements should be left blank to allow the pipe supplier and the installation contractor to establish the responsibilities for delivery.

Delivery of pipe shall begin at station (   ) on or before (   ), but not sooner than (   ) weeks after receipt of approved engineering details, and proceed at the rate of (   ) feet per week.

CHAPTER 14

Guide Specifications for Installation of Pipe

General

The procedures and methods for the installation of concrete pressure pipe may vary from one area to another and, in certain details, from one type of pipe to another. These variable factors are attributable to differences in types of soil, height of water tables, available bedding materials, climate, and local practices.

In all installations the basic objectives are the same—to obtain a workmanlike job within the established standards for design and safety that will ensure the successful performance of the pipeline. The following guide specification outlines the essential requirements for pipe installation.

Pipe Installation

General. This section of the specifications covers the excavation, bedding, laying, and backfilling of all pipe, fittings, and special pipe in the sizes and classes shown in the contract documents.

Excavation and subgrade. Excavation for pipe trenches shall be to the line, grade, and dimensions shown in the contract documents. In rock or other unyielding material, the trench shall be over-excavated by at least 3 in. below the established grade line and the over-excavation replaced with sand, gravel, or select excavated material as shown on the typical trench sections in the contract documents. When the bottom of a trench is unsuitable as a foundation for the pipe because of unstable material, the trench shall be over-excavated as directed by the engineer, and the over-excavation replaced with sand, gravel, or crushed rock.

Laying. The pipe shall be laid to the line and grade shown in the contract documents. The work shall be scheduled so that the bell ends of the pipe face in the direction of laying, wherever practicable. During the laying of the pipe, the pipe trench shall be kept free of water which might impair the integrity of the bedding and joining operations.

Joining. Before joining the spigot into the bell of the pipe just previously laid, the spigot groove, the rubber gasket, and the bell shall be thoroughly cleaned; then the spigot groove, the rubber gasket, and the initial 2 in. of the bell shall be lubricated with a soft, vegetable soap compound. The gasket shall be positioned in the spigot groove so that the rubber is distributed uniformly around the circumference. The pipe shall be joined together to provide proper space between abutting ends. After the joint is assembled, a thin metal feeler gage shall be inserted between the bell and the spigot to check the position of the rubber gasket around the complete circumference of the pipe.

Bedding. Pipe shall be bedded to the elevation above the bottom of the pipe shown on the typical trench section in the contract document. Bedding material may be selected excavated material tamped thoroughly under the haunches of the conduit or, at the option of the contractor, the bedding may consist of imported free-draining granular material densified by tamping or by consolidation with water. Consolidation by hydraulic methods may be used only if both the materials to be densified and the native soil in which the trench is excavated are free-draining.

Backfill. Backfill above the bedding to the original ground surface shall be placed so that the resultant density will be approximately equivalent to the density of the original material in which the trench was excavated. Free-draining materials can be consolidated using jets, vibrators, or nozzles. Cohesive materials shall be compacted using tamping or rolling equipment. Materials within 6 in. of the pipe shall be free from rocks or clods larger than 3 in. in diameter. Power-operated hauling or rolling equipment shall not be allowed to travel over the pipe unless 2–3 ft of densified backfill has been placed over the top of the pipe.

CHAPTER 15
Transportation of Pipe

General

Concrete pressure pipe is seldom made for stock purposes. For this reason, curing and storage yards of most manufacturers have space available for only a few weeks of storage after curing is completed. When ready, pipe can be delivered immediately to the job site. Prolonged storage in the manufacturer's yard or at a location at the job site is uneconomical.

Completed pipe must be stored vertically or horizontally on skids or on the ground, and can be carried by crane, forklift truck, or other suitable equipment. Certain types and sizes of pipe may sometimes be double decked or even triple decked when storage facilities are overtaxed. The manufacturer's advice, guidance, and approval should be requested before multiple-level storage is attempted by other than the manufacturer.

Selection of Transportation Mode

Concrete pressure pipe is generally transported from the manufacturer's yard to the area of installation by flatbed truck, flatbed trailer, low-bed trailer, or rail flatcar. Railroad piggyback and barge transportation have recently gained in importance. The method of hauling depends on comparative shipping costs primarily based on the distance involved, weight and size of pipe, availability of transporting equipment, and delivery conditions at the job site.

Unloading a typical flatbed truck trailer loaded with 36-in. concrete pressure pipe

Special double-drop highway truck trailer loaded with a single large-diameter pipe

Motor Truck Transportation

For distances up to approximately 300 miles, motor truck transportation is generally used because it is the most economical way to deliver pipe close enough to the point of installation. With this type of transportation, most manufacturers offer delivery to points along the trench that are accessible to trucks. Pushing or pulling is seldom permitted since this can easily damage a truck. Also, when hauling over excessively rough terrain, the erratic forward movement and twisting of the trailer can result in damage to pipe. Access to and departure from the delivery point should be specified in the terms of the purchase contract. Truck transportation is controlled by state highway restrictions with regard to weight, width, and height of loads. Generally, the tractor-trailer type of equipment is adaptable under most state regulations to provide for hauling loads up to 22 tons, 8 ft wide, 13½ ft above the road, and 40 ft long. The actual route to be traveled may further restrict these limitations if there are low overpasses, weak bridges, or narrow tunnels.

Special permits. Special permits exempting weight, height, and width limitations can be obtained from states if the vehicle provides for adequate weight distribution on the highways, and if the load will dimensionally clear obstructions along the route of movement.

Rail Transportation

Rail transportation is used when some of the following conditions exist: (1) shipping distances over 300 miles, (2) loads exceeding normal highway dimensions, and (3) delivery to a storage site served by railroad or to a trench site adjacent to a railroad line.

Transportation by railway is also limited by weight and clearance restrictions. Generally, 50-ft long flat cars are used for railway delivery. The normal capacity of flat cars is dependent on the size of pipe but ranges from 32 pieces of 16-in. diameter pipe to one piece of 120-in. or larger diameter pipe. Dimensional clearances are

TRANSPORTATION OF PIPE

Rail flatcars loaded with 144-in. diameter concrete pressure pipe

variable with usual maximum load widths of 10 ft 6 in. and load heights of 15 ft above the rail; however, exceptions to these limits may be procured from the originating railroad. Such clearances are especially important on large diameter pipe and fittings.

Railroad piggyback. Railroad piggyback provides some of the long distance economies of railroad transportation plus hauling of trucks to the delivery point. The pipe is loaded on trailers at the manufacturer's plant and the trailers are then pulled by tractor to the nearest railroad piggyback ramp where they are loaded onto special railroad flat cars.

The most common method of piggyback is Plan 2½, under which the railroad is only responsible for the pipe from the time the loaded trailer arrives at the origin railroad's piggyback trailer parking lot until the loaded trailer is available to the hauler at the destination railroad's piggyback trailer parking lot. During transportation, the railroads will only pay damage claims if they (the railroad company) are proven negligent. Arrangements for transporting the empty trailer from the railroad yard to the manufacturer's yard and for return of the loaded trailer to the railroad yard are usually made by the manufacturer. Transporting the loaded trailer from the destination railroad yard to the delivery point, returning the empty trailer to the railroad yard, and any applicable detention charges on the trailer are the responsibility of the manufacturer or purchaser, depending on the terms of delivery.

Piggyback transportation has about the same general weight and dimensional

Flatbed piggyback truck trailers loaded with small-diameter pipe

restrictions as motor trucks. The difference is that low-bed or cradle-type trailers are rarely available and over-dimension or over-weight shipments are seldom accepted by the railroads.

Barge Transportation

Barge transportation is usually limited to pipe that is too large for feasible rail shipment, delivery to a job site that can be reached by water only, or when the volume and distance produce economies compared to other modes of transportation. Barges range from harbor scows to ocean-going converted ships, and, therefore, the capacity is quite varied.

Loading Procedures

Loading procedures vary according to the size of pipe involved, the handling method used, and the type of vehicle to be loaded. Most pipe is shipped on flatbed vehicles and is loaded with forklifts if the weight per piece is under 65,000 lb. Loading larger pipe is generally accomplished with cranes. Cranes are also used when loading barges.

Close attention must be given to adequate blocking and bracing of pipe in conformity to rules prescribed by the Department of Transportation and the American Association of Railroads. Timbers are generally used for pipe cradles and separators. The pipe must be secured with chains, cable, or steel bands.

Delivery and Unloading

It is important that the manufacturer and purchaser agree on a definite delivery location prior to the shipment of pipe. The location may be f.o.b. carrier at the manufacturer's plant or at points of delivery. Trench-side delivery may specify unloading and stringing of pipe by the supplier, the contractor, or the manufacturer. In any event, the manufacturer is responsible for furnishing pipe in satisfactory condition in accordance with the delivery terms.

It is the purchaser's responsibility to inspect shipments at points of delivery. If the material is damaged, the purchaser must notify the carrier immediately and request a written damage inspection report. The manufacturer must also be notified

Specialized equipment for handling and transporting extremely large-diameter pipe. The pipe in this photo are 252 in. (I.D.) and weigh 225 tons each

of the damage. According to interstate commerce law, the consignee cannot refuse to accept a damaged shipment, but, if the damage is extensive, the carrier will usually agree not to unload it. Handling of pipe beyond the agreed point of delivery is the purchaser's responsibility.

Unloading equipment. The equipment for unloading concrete pressure pipe varies with the method of transportation and the facilities at the destination. Whether the pipe has to be unloaded at a siding, strung along the ditch line, or stockpiled should also be considered.

With few exceptions, cranes of the proper capacity using steel cable, belt slings, or specially designed devices can load or unload any size of pipe regardless of the mode of transportation. Forklift trucks and front-end loaders that have sufficient capacity can unload pipe from most carriers if the delivery location is accessible and if terrain conditions are suitable for operating the equipment. Front-end loaders should have the bucket replaced with forks or have forks installed on the lip of the bucket. A choke cable attached to the bucket is also frequently used. The uprights of the forks and the sides of the bucket should be cushioned with suitable material to prevent exterior damage to the pipe.

Special equipment can be designed for handling and assembling large diameter pipe; one example of such equipment is the "pipe-mobile."

Although concrete pressure pipe is an exceptionally rugged product, reasonable care should be exercised to prevent bumping and possible damage to pipe ends.

Approximately 18,000 ft of concrete pressure pipe has been loaded onto this ocean-going barge